178
Advances in Polymer Science

Advances in Polymer Science

Recently Published and Forthcoming Volumes

Microseparation of Block Copolymers II
Volume Editor: Abetz, V.
Vol. 190, 2005

Microseparation of Block Copolymers I
Volume Editor: Abetz, V.
Vol. 189, 2005

Intrinsic Molecular Mobility and Toughness of Polymers II
Volume Editor: Kausch, H. H.
Vol. 188, 2005

Intrinsic Molecular Mobility and Toughness of Polymers I
Volume Editor: Kausch, H. H.
Vol. 187, 2005

Polysaccharides I
Structure, Characterisation and Use
Volume Editor: Heinze, T. T.
Vol. 186, 2005

Advanced Computer Simulation Approaches for Soft Matter Science II
Volume Editors: Holm, C., Kremer, K.
Vol. 185, 2005

Crosslinking in Materials Science
Vol. 184, 2005

Phase Behavior of Polymer Blends
Volume Editor: Freed, K.
Vol. 183, 2005

Polymer Analysis. Polymer Theory
Vol. 182, 2005

Interphases and Mesophases in Polymer Crystallization II
Volume Editor: Allegra, G.
Vol. 181, 2005

Interphases and Mesophases in Polymer Crystallization I
Volume Editor: Allegra, G.
Vol. 180, 2005

Inorganic Polymeric Nanocomposites and Membranes
Vol. 179, 2005

Polymeric and Inorganic Fibres
Vol. 178, 2005

Poly(arylene ethynylene)s
From Synthesis to Application
Volume Editor: Weder, C.
Vol. 177, 2005

Metathesis Polymerization
Volume Editor: Buchmeiser, M.
Vol. 176, 2005

Polymer Particles
Volume Editor: Okubo, M.
Vol. 175, 2005

Neutron Spin Echo in Polymer Systems
Authors: Richter, D., Monkenbusch, M., Arbe, A., Colmenero, J.
Vol. 174, 2005

Advanced Computer Simulation Approaches for Soft Matter Sciences I
Volume Editors: Holm, C., Kremer, K.
Vol. 173, 2005

Microlithography · Molecular Imprinting
Vol. 172, 2005

Polymer Synthesis
Vol. 171, 2004

NMR · 3D Analysis · Photopolymerization
Vol. 170, 2004

Long-Term Properties of Polyolefins
Volume Editor: Albertsson, A.-C.
Vol. 169, 2004

Polymers and Light
Volume Editor: Lippert, T. K.
Vol. 168, 2004

Polymeric and Inorganic Fibers

With contributions by
J. J. M. Baltussen · P. den Decker · T. Ishikawa · M. G. Northolt ·
S. J. Picken · R. Schlatmann

Springer

The series presents critical reviews of the present and future trends in polymer and biopolymer science including chemistry, physical chemistry, physics and material science. It is addressed to all scientists at universities and in industry who wish to keep abreast of advances in the topics covered.

As a rule, contributions are specially commissioned. The editors and publishers will, however, always be pleased to receive suggestions and supplementary information. Papers are accepted for "Advances in Polymer Science" in English.

In references Advances in Polymer Science is abbreviated *Adv Polym Sci* and is cited as a journal.

The electronic content of APS may be found springerlink.com

Library of Congress Control Number: 2004117780

ISSN 0065-3195
ISBN-10 3-540-24016-0 **Springer Berlin Heidelberg New York**
ISBN-13 978-3-540-24016-7 **Springer Berlin Heidelberg New York**
DOI 10.1007/b104206

Springer is a part of Springer Science+Business Media
springeronline.com

Printed in Germany

Cover design: KünkelLopka GmbH, Heidelberg/design & production GmbH, Heidelberg
Typesetting: Fotosatz-Service Köhler GmbH, Würzburg

Printed on acid-free paper 02/3141 xv – 5 4 3 2 1 0

Editorial Board

Advances in Polymer Science
Also Available Electronically

For all customers who have a standing order to Advances in Polymer Science, we offer the electronic version via SpringerLink free of charge. Please contact your librarian who can receive a password for free access to the full articles by registering at:

springerlink.com

If you do not have a subscription, you can still view the tables of contents of the volumes and the abstract of each article by going to the SpringerLink Homepage, clicking on "Browse by Online Libraries", then "Chemical Sciences", and finally choose Advances in Polymer Science.

You will find information about the

- Editorial Board
- Aims and Scope
- Instructions for Authors
- Sample Contribution

at springeronline.com using the search function.

Contents

Adv Polym Sci (2005) 178: 1–108
DOI 10.1007/b104207

The Tensile Strength of Polymer Fibres

M. G. Northolt (✉) · P. den Decker · S. J. Picken · J. J. M. Baltussen · R. Schlatmann

[1] Magellan Systems International, 8310 Shell Road, Richmond VA 23237, USA
mafe.northolt@wxs.nl

[2] Teijin Twaron Research, P.O. Box 9600, 6800 TC Arnhem, The Netherlands

[3] Delft University of Technology, Polymer Materials & Engineering, Julianalaan 136, 2628 BL Delft, The Netherlands
s.j.picken@tnw.tudelft.nl

[4] Akzo Nobel Chemicals Research Arnhem, P.O. Box 9300, 6800 SB Arnhem, The Netherlands

Abstract A theory of the tensile strength of oriented polymer fibres is presented. From an analysis of the observed fracture envelope it is shown that the criterion for fracture of the fibre is either a critical shear stress or a critical shear strain. Owing to the chain orientation distribution in the fibre, the initiation of fracture is likely to occur in domains whose symmetry axes have orientation angles in the tail of this distribution. By considering the fibre as a molecular composite, the tensile strength is calculated as a function of the modulus. The results are compared to the observed values of PET, POK, cellulose II, PpPTA, PBO and PIPD fibres. In addition, the relation between the ultimate strength and the chain length distribution is investigated. By using the critical shear strain as a fracture criterion in the Eyring reduced time model, relations are derived for the fibre strength as a function of the load rate, as well as for the lifetime under constant load. Moreover, this model predicts the dependence of the strength on the temperature. The theoretical relations are compared to the experimental results on PpPTA fibres.

Keywords Polymer fibre · Strength · Chain length distribution · Creep fracture · Lifetime · Poly(*p*-phenylene terephthalamide)

Abbreviations and Symbols

A	Cross-sectional area
c	Concentration
d_c	Interplanar spacing
d.r.	Draw ratio
D	Diameter of the fibre
DABT	Poly(*p*-benzanilide terephthalamide)
DP	Degree of polymerisation
e_c	Chain modulus
e_1	Modulus transverse to the chain axis
esd	Estimated standard deviation
E	Fibre modulus
ERT	Eyring reduced time
$f(z)$	Chain length distribution
$f_w(z)$	Molecular weight distribution
g	Shear modulus of the domain
g_v	Apparent shear modulus
G	Torsional modulus of the filament

HT	High tenacity
$h(z)$	Crossing length distribution
$I(U)$	Transition density distribution
$j(t)$	Creep compliance
k_B	Boltzmann constant
K	Kelvin
L_d	Contour projection length of the chain
L_C	Contour length
L_G	Griffith crack length
L_P	Persistence length of the chain
m	Weibull modulus
M_n	Number-average molecular weight
M_w	Weight-average molecular weight
M_z	Z-average molecular weight
m.u.	Monomeric unit
N_A	Avogrado's number
p	Distance between periodic force centres
PAN	Polyacrylonitrile
PBO	Poly(*p*-phenylene benzobisoxazole)
PE	Polyethylene
PET	Poly(*p*-ethylene terephthalate)
PIPD	Poly({2,6-diimidazole[4,5-*b*:4′,5′-*e*]pyridinylene-1,4(2,5-dihydroxy)phenylene})
POK	Polyetherketone
PpPTA	Poly(*p*-phenylene terephthalamide)
$P(\sigma)$	Cumulative failure probability
$\langle P_2 \rangle$	Internal order parameter
P_D	Order parameter of the directors
q	Crack size
r	Radius of chain cross section
RH	Relative humidity
s.s.	Spinning speed
t	Time
t_b	Lifetime
T	Temperature
T_g	Glass transition temperature
T_{ni}	Nematic–isotropic transition temperature
T_0	Reference or Vogel temperature
u	Chain length
u_a	Average chain length
u_c	Bonded chain length
u_0	Monomer length
U	Activation energy
UHMW	Ultra-high molecular weight
V	Volume
V_c	Chain volume fraction
V_{cell}	Unit cell volume
W	Strain energy
W_a	Activation energy of creep
W_b	Fracture energy
W^C	Strain energy of the chain
W^S	Shear energy

W^S_m	Maximum shear energy
W^S_0	Shear energy of fracture
w_γ	Surface energy of a crack
z	Chain length in monomeric units
z_c	Bonded chain length in m.u.
z_n	Number-average chain length in m.u.
z_w	Weight-average chain length in m.u.

Greek symbols

β	Critical shear strain in tensor notation
γ	Shear strain in engineering notation
δ	Relaxation time
ε	Strain
ε_b	Strain at fracture
ε_b^s	Shear strain at fracture
ε_f	Fibre strain
ε_0	Ultimate strain at fracture
ε_f^y	Yield strain of the fibre
ε_{13}	Shear strain in tensor notation
ε_{13}^v	Viscoelastic shear strain of a domain
ε_{13}^y	Shear yield strain in tensor notation
ζ	Strength of orienting nematic potential
η	Viscosity
θ	Orientation angle at stress σ
θ_b	Orientation angle at fracture
Θ	Orientation angle in the unloaded state
λ	Load rate
ν	Frequency
$\rho(\theta)$	Orientation distribution of the chains
σ	Stress
σ_b	Tensile strength
σ_b^s	Fibre strength based on shear deformation only
σ_{comp}	Strength of a macrocomposite
σ_0	Ultimate strength
σ_L	Longitudinal strength
σ_T	Transverse strength
σ_y	Yield stress
τ	Shear stress
τ_b	Shear strength
τ_m	Maximum shear stress
τ_n	Normalised shear stress
τ_y	Shear yield stress
τ_0	Ultimate shear strength
χ	Euler's constant
ω	Angular frequency
Ω	Activation volume

1
Introduction

Organic polymer fibres offer an impressive range of mechanical properties. The tensile modulus of these fibres varies between 5 and 330 GPa, with a tensile strength up to 7 GPa, a compressive strength up to 1.7 GPa, and a temperature resistance up to 400 °C. The tensile curves of these fibres for temperatures below the glass transition temperature, including the yield phenomenon, are well described by the continuous chain model [1–10]. Considerable attention has been given in the literature to the relation between the tensile strength and the chain length distribution [11–14]. As will be shown here, there are also other factors of similar importance which determine the strength of a polymer fibre. In this report a relationship is derived describing the fibre strength as a function of the orientation distribution of the chains and the intrinsic mechanical properties, such as the elastic modulus of the polymer chain and the modulus for shear between the chains. In addition, a modified version of Yoon's model for the description of the relation between the strength and the chain length distribution is presented. Finally, a model is proposed for the dependence of the fibre strength on the time and the temperature.

Before embarking on the discussion of these intrinsic factors determining the strength of polymer fibres, the effect of structural and morphological imperfections on the fibre strength are briefly discussed. During the manufacturing process of polymer fibres all kinds of imperfections are introduced, like structural inhomogeneities, impurities and voids. These so-called extrinsic factors result in an imperfect bonding between the chains and may give rise to stress concentrations, which after a catastrophic growth of pre-existing cracks can lead to fracture. These imperfections cause the size effects, viz. the transverse effect or the dependence of the strength on the fibre diameter, and the longitudinal effect or the dependence of the strength on the test length [15–17]. Two different approaches can be recognised for the description of the size effects. The first is based on Griffith's theory of crack propagation, which considers the energy balance between the external work, the surface energy of the crack and the elastic energy of the material [18, 19]. This theory is based on the elastic theory of infinitesimal deformations, and so does not apply to highly deformable materials. It can be applied to the transverse effects and leads to the semi-empirical equation for the strength of a material

$$\frac{1}{\sigma_b} = \frac{1}{\sigma_0} + K \cdot \sqrt{D} \tag{1}$$

where σ_b is the actual strength of the fibre, σ_0 the strength of the flawless fibre or the ultimate strength, K a constant and D the diameter of the fibre [20]. It was later shown by Penning et al. that the scaling of the tensile strength with $D^{-0.5}$ can be derived from geometrical considerations as well [16]. An example of this relation is presented in Fig. 1, where the yarn strength of poly(*p*-phenyl-

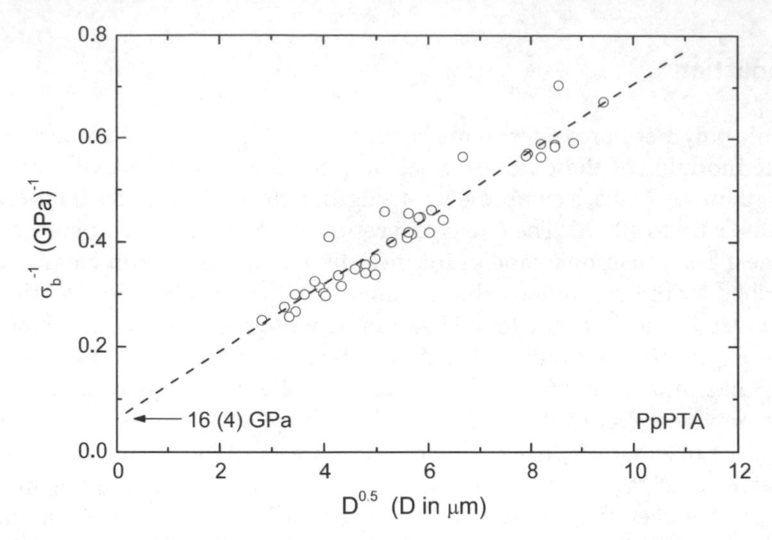

Fig. 1 The inverse of the observed strength of PpPTA yarns versus the square root of the diameter of the filaments. Linear regression yields σ_b^{-1}=0.063(16)+0.0643(27) $D^{0.5}$ (GPa)$^{-1}$ with σ_0=16(4) GPa, and estimated standard deviations in parentheses

ene terephthalamide) or PpPTA is plotted versus the filament diameter [21]. Apparently, for the strength of a flawless PpPTA filament the extrapolation yields σ_0=16±4 GPa. As will be shown in this report, this value is too large.

Whereas in the second approach of the size effects it is also assumed that fracture is controlled by defects, the strength is now considered a statistically distributed parameter rather than a physical property characterised by a single value. The statistical distribution of fibre strength is usually described by the Weibull model [22, 23]. In this weakest-link model the strength distribution of a series arrangement of units of length L_0 is given by

$$P(\sigma) = 1 - \exp\left[-\frac{L}{L_0}\left(\frac{\sigma}{\sigma_p}\right)^m\right] \tag{2}$$

where $P(\sigma)$ is the cumulative failure probability at a stress σ, σ_p a scaling parameter and m the Weibull modulus. To make a so-called Weibull plot of a yarn $P(\sigma)$ is approximated by

$$P = \frac{n_i}{n+1} \tag{3}$$

where n_i is the number of filaments that have broken at or below a stress σ and n is the total number of filaments tested. The length dependence is expressed through the test length L and can be written as

$$\log[-\log(1-P)] - \log L + \log L_0 = m\log\sigma - m\log\sigma_p \tag{4}$$

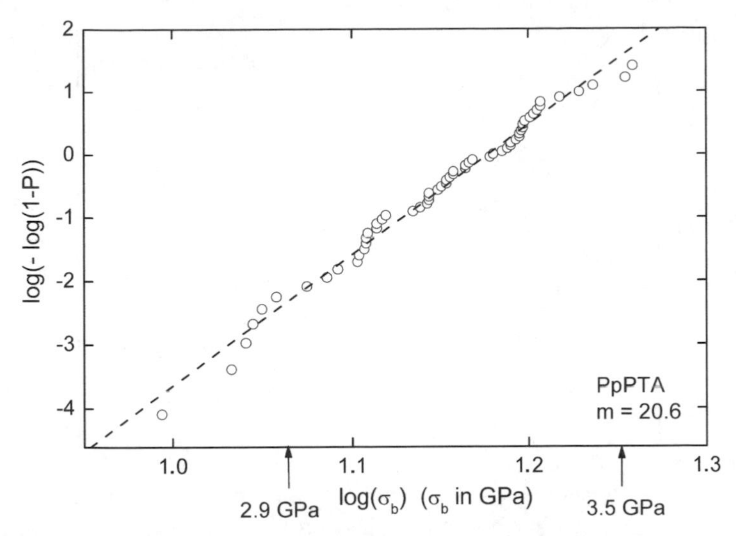

Fig. 2 Weibull plot of the filament strength for a test length of 10 cm of a PpPTA yarn yielding a Weibull modulus of 20.6. The average filament strength is 3.2 GPa

Thus, given a Weibull distribution of the filament strength, a plot of log [–log(1–P)] versus logσ results in a straight line with a slope m. For the range $5<m<30$ the relation between the coefficient of variance (cv) of the filament tenacity distribution and m is given by cv=1.2 m^{-1}. Figure 2 presents an example of a Weibull plot of the filament strength of a PpPTA yarn, yielding a Weibull modulus of 20.6.

The average fracture stress of the filaments for a test length L is given by

$$\langle\sigma\rangle = \sigma_p L^{-\frac{1}{m}} \Gamma\left(1 + \frac{1}{m}\right) \tag{5}$$

where Γ is the gamma function [15]. Equation 5 shows that the average strength depends on the test length of the fibre sample, which can be approximated by

$$\log(\langle\sigma\rangle) \approx C - \frac{1}{m}\log(L) \quad \text{with} \quad C = \log[\sigma_p \Gamma(1 + 1/m)] \tag{6}$$

Thus, the Weibull modulus can be derived from the strength distribution at a fixed test length as shown by Eq. 4, as well as from a plot of the average filament strength as a function of the test length according to Eq. 6. In Fig. 3, an example of the relation in Eq. 6 is presented for a PAN-based carbon fibre [8]. From this plot a value m=7.2 with an estimated standard deviation (esd) of 0.7 is derived, whereas the m values obtained from the strength distributions at fixed length are 5.2 (0.6) for 2 mm, 5.1 (0.6) for 10 mm and 4.6 (0.6) for 25 mm, with esd values in parentheses. Apparently the length effect is weaker than expected

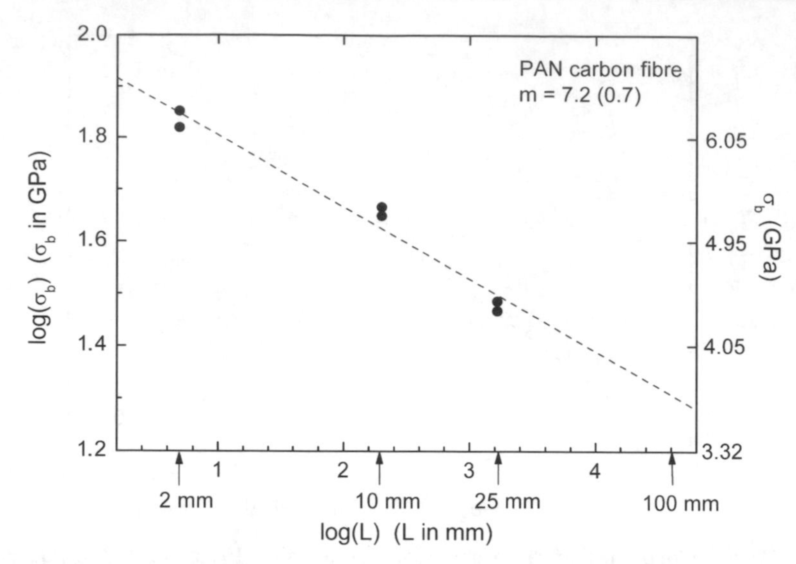

Fig. 3 The natural logarithm of the average filament strength (n=40) as a function of the natural logarithm of the test length for an intermediate-modulus, PAN-based carbon fibre with an impregnated bundle strength of 5.7 GPa [8]

from the width of the strength distributions at fixed length, which may indicate that adjacent segments in the carbon filament are not statistically independent, as is assumed by the weakest-link theory, i.e. the model is not quite suitable. With regard to the application of fibres in uniaxially reinforced composites, the critical length of a fibre is the test length for which the average filament tenacity is equal to the impregnated bundle strength. Since the impregnated bundle strength of this carbon fibre is 5.7 GPa, it follows from Fig. 3 that the critical length is 4.5 mm. Weibull moduli of filaments taken from yarns range from 5 for brittle carbon fibres to about 50 for ductile melt-spun poly(p-ethylene terephthalate) (PET) fibres.

Penning et al. studied the transverse and longitudinal size effects in high-strength ultra-high molecular weight (UHMW) polyethylene (PE) fibres and found that the length or longitudinal size effects become weaker as the tensile modulus of the fibre increases, whereas the transverse effect becomes more pronounced as the modulus increases [16]. In particular, the length effect disappeared almost completely for PE fibres with a draw ratio of 70. This was attributed by Penning et al. to the fact that the high-modulus PE fibres do not possess a distribution of macroscopic flaws, occurring at distances of the same order of magnitude as the applied test lengths, but contain a microscopic defect structure at very short intervals of about 100 nm. They concluded that, apparently, transverse and longitudinal effects have different physical backgrounds and, therefore, cannot be described simultaneously by statistical theories such as the weakest-link hypothesis. In the case of high-modulus/high-strength fibres, such as PpPTA and poly(*p*-phenylene benzobisoxazole) or PBO, and poly({2,6-diimidazole[4,5-*b*:4′,5′-*e*]pyridinylene-1,4(2,5-dihydroxy)phenylene}) or PIPD, made by the wet-spinning process, the transverse size effect is difficult to detect, because a decrease of the filament diameter is often accompanied by an increase of the tensile modulus. As will be shown in Sect. 2 this results in an increase of the strength. In this regard the conclusions drawn from Fig. 1 should be considered with some caution. With regard to the observation of Penning at el. that the longitudinal size effect becomes weaker as the modulus increases, it will be shown in this report that, by applying Griffith's theory on cracks in anisotropic fibres, elongated cracks are supposed to be more damaging in low-oriented fibres than in highly oriented fibres.

An extensive discussion of the concept of fibre strength, the Weibull modulus, and its relation to fracture toughness has been given by Van der Zwaag [15]. An increase of the material toughness will result in an increase of the Weibull modulus, because incorporation of local plastic deformation will decrease the stress concentrations in the fibre. This is much more easily achieved with flexible polymer chains than with rigid-rod chains, not to mention the graphitic planes in carbon fibres. Van der Waals and hydrogen bonds offer more advantages in this respect than covalent bonds between the building elements of the fibre. This explains the observation that the Weibull modulus of yarns decreases according to the sequence: PE, PET, cellulose, PpPTA, PBO and carbon fibre. The particularly low value of the Weibull modulus of carbon fibres is a consequence of the brittleness of these fibres. Therefore, the increase of the strength of carbon fibres calls for extreme care at each stage of the process to preclude any kind of flaw-producing impurity [24].

Melt-spun fibres such as PET displaying a "flag" or a plastic mode of deformation at the end of the tensile curve show a large variation of the elongation at break. At low tensile speeds these fibres display ductile fracture initiated by crack growth, and for increasing testing speeds the melt fracture morphology becomes dominant. Adiabatic heating of the fibre during rapid cold drawing will raise the temperature well above the glass transition temperature [25]. But even at medium strain rates of 100% per minute tiny irregularities in the fibre may cause localised drawing or "necking", whereby the temperature can approach the melting temperature, resulting in an extra elongation before fracture. Hence, this random phenomenon of "hot spots" occurring during cold drawing causes the wide range of elongations at break observed during filament testing of PET fibres. With regard to the failure mode, it is significant to note that polymer fibres without a melting temperature, such as cellulose, PpPTA, PBO and PIPD, generally display a more or less fibrillar fracture morphology. This is in contrast to polymer fibres having a melting temperature like PE, PET and the aliphatic polyamides, which often show melt-flow phenomena during cold drawing.

Weibull plots of various fibre properties, such as the filament count, modulus, elongation at break and the strength, can provide important information on the quality and performance of the manufacturing process. The results can be used to formulate a strategy for the improvement of the yarn properties.

As will be shown in this report, polymer fibres gain additional strength by an increase of the molecular weight and by a more contracted orientation distribution, i.e. a higher modulus. For the wet-spun fibres, a strength increase can be achieved by improvement of the coagulation process, which makes for a more uniform structure and chain orientation in the cross section of the fibre, and by a reduction of the amount of impurities.

For an understanding of the fracture process and the dependence of the strength on the chain orientation distribution and the basic elastic constants, we briefly discuss the tensile deformation of polymer fibres. The continuous chain model provides a good description of the tensile curve of a polymer fibre [1–10]. In this model the fibre is built up of parallel oriented fibrils with equal properties. Thus it is assumed that a mechanical model of the extension of a single fibril as a function of the fibre stress gives a complete description of the tensile deformation of the fibre. Each fibril is a series arrangement of domains consisting of perfectly oriented chains. The domains are cylindrically symmetric around the chain axis and the axes of the domains follow an orientation distribution, $\rho(\Theta)$, in the unloaded state. The elastic constants of the domain most relevant to the tensile extension of the fibre are the chain modulus, e_c, and the modulus for shear between adjacent chains, g. Figure 4 shows the stresses acting on a domain due to a tensile stress on the fibre and Figs. 5 and 6 depict schematic representations of the domain deformation according to the continuous chain model. The fibre strain is given by

$$\varepsilon_f = \frac{\sigma\langle\cos^2\theta\rangle}{e_c} + \frac{\langle\cos\theta\rangle - \langle\cos\Theta\rangle}{\langle\cos\Theta\rangle} \tag{7}$$

where Θ is the initial orientation angle of the chain axis at zero load and θ the angle at a tensile stress σ. The averaging is performed over the chain orientation distributions $\rho(\Theta)$ and $\rho(\theta)$ of the domains in the fibril. As shown by Eq. 7 the fibre strain is composed of two contributions, viz. the elastic chain exten-

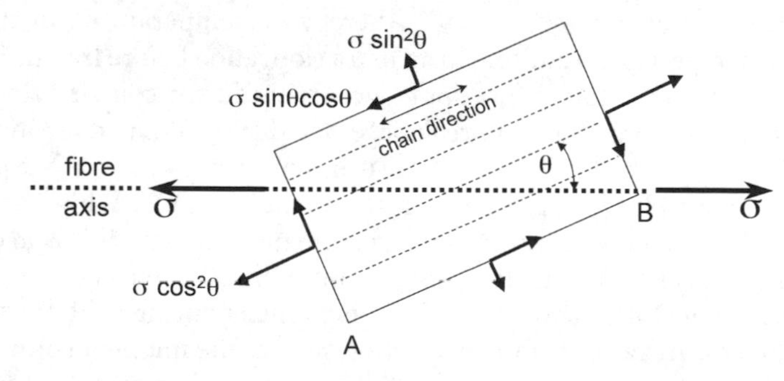

Fig. 4 The four normal stresses and the four equal shear stresses acting on the domain in the fibre under a tensile stress σ. The chains are parallel to AB and make an angle θ with the fibre axis

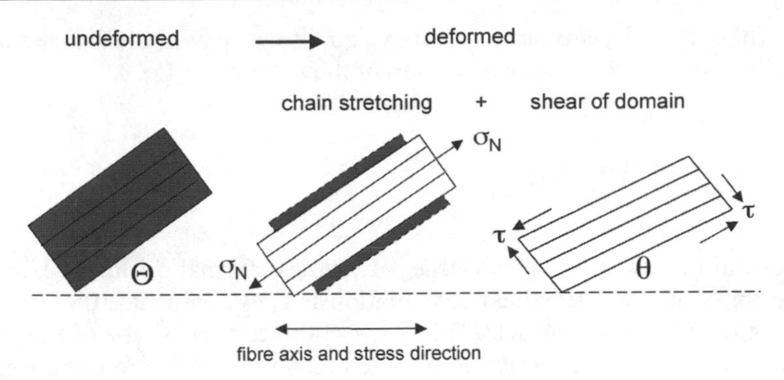

Fig. 5 Schematic representation of the domain contributions to the tensile deformation of the fibre: chain stretching and chain rotation due to shear deformation

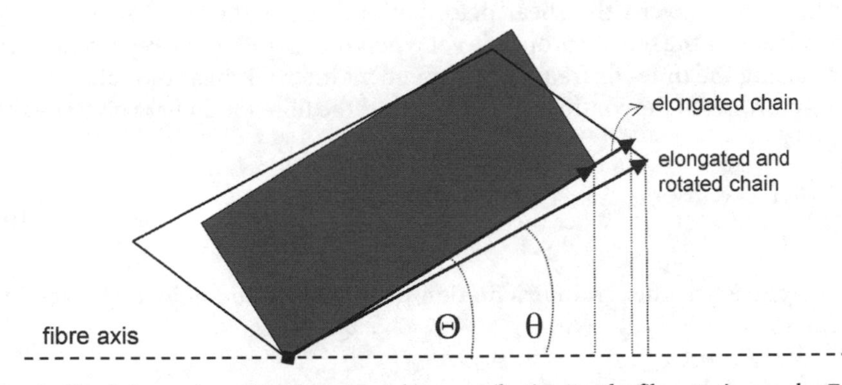

Fig. 6 The deformation of the domain and its contributions to the fibre strain, see also Eq. 7

sion brought about by the normal stress $\sigma\cos^2\theta$ and the change of the projected chain length due the rotation of the chain axes towards the fibril axis. This rotation is caused by the resolved shear stresses $|\tau|=\sigma\sin\theta\cos\theta$ acting on the domains. The key equation for the elastic shear deformation of a domain in a fibre is

$$\tan(\theta-\Theta) = -\frac{\sigma}{2g}\sin\theta\cos\theta \tag{8}$$

In this equation $\tau=-\sigma\sin\theta\cos\theta$ is the shear stress acting on a domain in the fibre and $\tan(\theta-\Theta)\approx-(\Theta-\theta)$ is the shear strain ε_{13} in tensor notation. Using the engineering notation for the shear strain $\gamma=2\varepsilon_{13}$, Eq. 8 becomes $\tau=g\gamma$ and represents Hooke's equation for shear deformation. Note that during tensile extension of the fibre the shear stress has by definition a negative value. The shear modulus g is also called the internal shear modulus in order to distinguish it from the shear or torsion modulus, G, of the filament itself. For a polymer fibre without cracks G is approximately equal to g. We will often make use of the following analytical approximation of the solution of Eq. 8:

$$\tan\theta = \frac{\tan\Theta}{\left(1+\dfrac{\sigma}{2g}\right)} \tag{9}$$

In the continuous chain model, a large part of the deformation during extension of the fibre consists of the shear deformation as a result of which the chain orientation distribution contracts. This leads to the concave shape of the tensile curve, often found for polymer fibres. Therefore, this description of the extension of the fibre implies a strain hardening process.

Chain stretching is governed by the covalent bonds in the chain and is therefore considered a purely elastic deformation, whereas the intermolecular secondary bonds govern the shear deformation. Hence, the time or frequency dependency of the tensile properties of a polymer fibre can be represented by introducing the time- or frequency-dependent internal shear modulus $g(t)$ or $g(\nu)$. According to the continuous chain model the fibre modulus is given by the formula

$$\frac{1}{E} = \frac{1}{e_c} + \frac{\langle\sin^2\Theta\rangle_E}{2g} \tag{10}$$

where $\langle\sin^2\Theta\rangle_E$ is the strain orientation parameter in the unloaded state defined as

$$\langle\sin^2\Theta\rangle_E = \frac{\int_0^{\pi/2} \sin^2\Theta\cos\Theta\rho(\Theta)\sin\Theta d\Theta}{\int_0^{\pi/2} \cos\Theta\rho(\Theta)\sin\Theta d\Theta} \tag{11}$$

The function $N(\Theta)=\rho(\Theta)\sin\Theta$ is proportional to the total number of domains in the fibre at an angle Θ with the fibre axis. The determination of the modulus using sonic frequencies will yield a higher value of g than the method by which the modulus is derived from the initial slope of the tensile curve. For random orientation of the chains in a fibre $\langle\sin^2\Theta\rangle_E=0.5$ and the modulus of an isotropic fibre is given by $E_{iso}\approx 4g$.

The continuous chain model includes a description of the yielding phenomenon that occurs in the tensile curve of polymer fibres between a strain of 0.005 and 0.025 [1]. Up to the yield point the fibre extension is practically elastic. For larger strains, the extension is composed of an elastic, viscoelastic and plastic contribution. The yield of the tensile curve is explained by a simple yield mechanism based on Schmid's law for shear deformation of the domains. This law states that, for an anisotropic material, plastic deformation starts at a critical value of the resolved shear stress, $|\tau_y|=fg$, along a slip plane. It has been shown that the yield strain of a polymer fibre, ε_f^y, is a function of the fibre modulus or the chain orientation distribution and is given by

$$\varepsilon_f^y \approx \left(\frac{1}{e_c} + \frac{\langle \sin^2\Theta \rangle_E}{2g}\right) \frac{fg}{\sin\Theta_a \cos\Theta_a} \tag{12}$$

where Θ_a is the average angle of the distribution $\rho(\Theta)$. Except for highly oriented fibres Eq. 12 can be approximated by

$$\varepsilon_f^y \approx \frac{f \tan\Theta_a}{2} \tag{13}$$

In isotropic samples yielding begins in the domains with an angle $\Theta=\pi/4$ and Eq. 13 reduces to

$$\varepsilon_f^y \approx \frac{f}{2} \tag{14}$$

Indeed, it has been observed that the onset of yielding of isotropic polymers is approximately constant, $0.02<\varepsilon_f^y<0.025$, which implies that $0.04<f<0.05$ [1]. Above the shear yield strain, the plastic shear deformation of the domain satisfies a plastic shear law. For temperatures below the glass transition temperature, the continuous chain model enables the calculation of the tensile curve of a polymer fibre up to about 10% strain [6]. Figure 7 shows the observed stress–strain curves of PpPTA fibres with different moduli compared to the calculated curves.

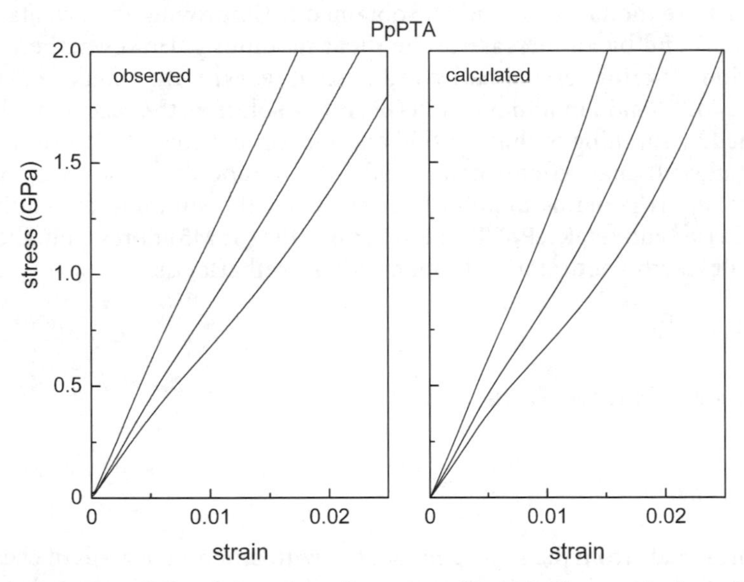

Fig. 7 Comparison of the observed tensile curves of PpPTA fibres with three different moduli with the curves calculated with the continuous chain model [6]

In a further development of the continuous chain model it has been shown that the viscoelastic and plastic behaviour, as manifested by the yielding phenomenon, creep and stress relaxation, can be satisfactorily described by the Eyring reduced time (ERT) model [10]. Creep in polymer fibres is brought about by the time-dependent shear deformation, resulting in a mutual displacement of adjacent chains [7–10]. As will be shown in Sect. 4, this process can be described by activated shear transitions with a distribution of activation energies. The ERT model will be used to derive the relationship that describes the strength of a polymer fibre as a function of the time and the temperature.

In order to simplify the discussion and to keep the derivation of the formulae tractable, the major part of this analysis is limited to a polymer fibre with a single orientation angle Θ. This angle is assumed to be a kind of average angle and a characteristic parameter of the orientation distribution of the chain axes.

For a proper understanding of the tensile deformation of a polymer fibre it is useful to know the approximate values of the various quantities. An imaginary PpPTA fibre is considered with a single orientation angle of Θ=9.6° that is loaded to a stress of 4 GPa. Typical values for PpPTA are e_c=240 GPa, g= 1.8 GPa and an orientation parameter of $\sin^2\Theta$=0.028 resulting in a modulus of 84 GPa. Using the continuous chain model, it can be calculated that the elastic fibre strain is composed of 0.016 due to chain stretching and of 0.011 due to chain rotation. The orientation angle at 4 GPa is θ=4.6°, which means a rotation angle of 5° or a shear strain in engineering units of $\gamma=2\varepsilon_{13}$ of 0.175 radians. So, a considerable fraction of the fibre strain is caused by the contraction of the chain orientation distribution, which increases for decreasing fibre modulus.

As the chain modulus of a polymer cannot be altered in a spinning process, a larger fibre modulus can only be obtained by improving the orientation of the chains and by an increase of the shear modulus g. However, there is one exception. After dissolving native cellulose fibres with the cellulose I conformation and a chain modulus of 138 GPa into a solution, the regenerated fibres obtained by spinning of this solution and subsequent coagulation always have the cellulose II chain conformation with a chain modulus of 88 GPa [26].

This report contains unpublished results on the spinning of cellulose II, polyetherketone (POK), PpPTA, DABT and PIPD (or M5) fibres from the Akzo Nobel Research Laboratories Arnhem in The Netherlands.

2 Fracture of a Polymer Fibre

2.1 Fracture Envelope

For fibres made from the same polymer but with different degrees of chain orientation the end points of the tensile curves, $\{\varepsilon_b, \sigma_b\}$, are approximately located on a hyperbola. Typical examples of this fracture envelope are shown in Figs. 8

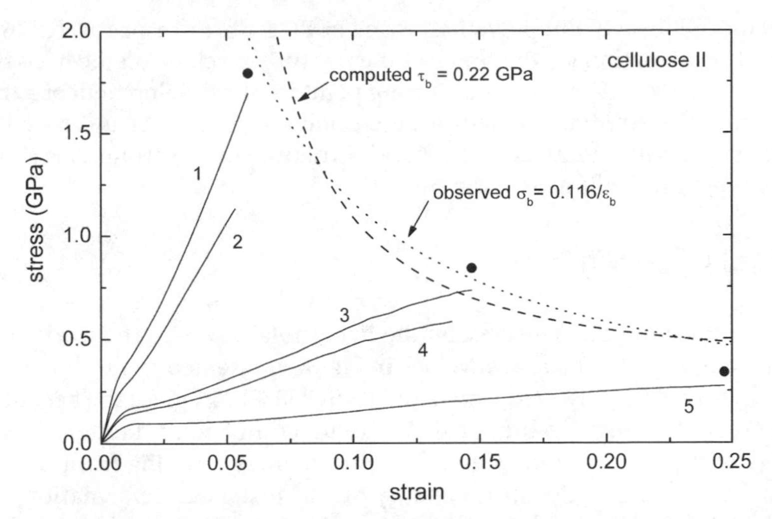

Fig. 8 Tensile curves of cellulose II fibres measured at an RH of 65%: (1) Fibre B, (2) Cordenka EHM yarn, (3) Cordenka 700 tyre yarn, (4) Cordenka 660 tyre yarn and (5) Enka viscose textile yarn [26]. The *solid circles* represent the strength corrected for the reduced cross section at fracture. The *dotted curve* is the hyperbola fitted to the end points of the tensile curves 1, 3 and 5. The *dashed curve* is the fracture envelope calculated with Eqs. 9, 23 and 24 using a critical shear stress τ_b=0.22 GPa

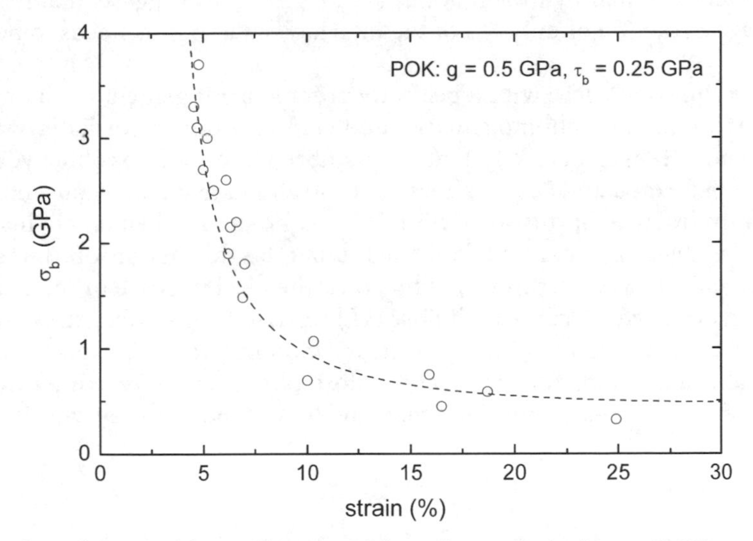

Fig. 9 The end points of the tensile curves of the various polyetherketone (POK) yarns spun by B.J. Lommerts [27–29]. The *dashed curve* has been calculated with Eqs. 9, 23 and 24 using τ_b=0.25 GPa

and 9 for cellulose II and polyetherketone or POK fibres, respectively [26–29]. A simple explanation for the shape of the fracture envelope on the basis of the cellulose II data will be used as a starting point for the development of a theory describing the strength of well-oriented polymer fibres. Assuming a linear tensile curve with modulus E, the work of fracture or the strain energy up to fracture per unit volume is given by

$$W_b = \int_0^{\varepsilon_b} \sigma d\varepsilon = \frac{1}{2} E\varepsilon_b^2 = \frac{1}{2}\sigma_b\varepsilon_b \tag{15}$$

Thus for the end points located on the hyperbola, $\sigma_b \propto 1/\varepsilon_b$, the work of fracture is constant. The fracture envelope in Fig. 8 represented by the dotted curve is the hyperbola $\sigma_b=2W_b/\varepsilon_b$, which for $W_b=0.058$ GJ m^{-3} gives the best fit with the observed strength values [26]. As discussed in Sect. 1, the tensile strain energy of the fibre is composed of the strain energy arising from the chain extension W^C and of the strain energy due to the shear deformation of the domains W^S. Cellulose II has a chain modulus e_c=88 GPa, which yields for the strain energy of the chain up to fracture of the fibre with the largest modulus (Fibre 1 in Fig. 8) $W_b^C=1/2\sigma_b^2/e_c=0.017$ GJ m^{-3}. This implies that the major part of the strain energy of the cellulose fibres is stored in the straining of the intermolecular hydrogen bonds as a result of the shear deformation between the chains. The shear strain energy up to fracture $W_b{}^S$ is about 7% of the total energy content of the hydrogen bonds in cellulose II, assuming that all possible intermolecular hydrogen bonds are formed. This suggests that for fibre breakage a fraction of only 7% of the intermolecular bonds needs to be broken.

In a cellulose II fibre with a perfectly ordered arrangement of chains and without impurities, inhomogeneities and voids, the only disorder is near the chain ends. Hence, it is likely that in this fibre almost all possible hydrogen bonds are formed and severe stress concentrators are virtually absent. Consequently, fracture of this ideal fibre is brought about when all chains have been dislodged or separated from each other. By this definition the tensile stress required to break this ideal fibre, σ_0, is the ideal strength or the ultimate strength of a perfect ordered cellulose II fibre. Assuming a linear stress–strain curve, we can estimate σ_0 for an arbitrary modulus value. Because the total intermolecular energy of the ideal cellulose fibre is about ten times the observed fracture energy, the maximum or total debonding energy is given by

$$W_m \approx 10W_b \tag{16}$$

For the total debonding or fracture energy of a perfect fibre with an elongation at break ε_0 and strength σ_0 we can write

$$W_m = \frac{1}{2}\sigma_0\varepsilon_0 \tag{17}$$

Assuming $\varepsilon_b = \sigma_b / E$ and $\varepsilon_0 = \sigma_0 / E$, Eqs. 15, 16 and 17 give for the ultimate strength of the perfect cellulose II fibre

$$\sigma_0 \approx \sqrt{10} \cdot \sigma_b \tag{18}$$

We take as an example the mechanical properties of a high-modulus and high-strength cellulose II fibre (Fibre 1 in Fig. 8) with a modulus of 45 GPa and a filament strength of 1.7 GPa [26]. According to the estimate of Eq. 18, the ultimate strength of an ideal cellulose II fibre with a modulus of 45 GPa is 5.4 GPa, which is 12% of the fibre modulus. This fraction of the modulus is in agreement with the estimates of the ultimate strength values of other materials [19]. However, it will be shown that this approach for the estimate of the ideal strength of a polymer fibre is too simple.

2.2 Critical Shear Stress and Critical Shear Strain

The presented explanation for the existence of the fracture envelope will be used in formulating a fracture criterion for polymer fibres. Let us suppose a hypothetical polymer fibre with chains having a single orientation angle Θ in the unloaded state. The shape of the fracture envelope is now calculated by taking into account the shear deformation of the chains only. For this case the work per unit volume up to fracture is given by

$$W_b^S = \int_0^{\gamma_b} \tau d\gamma \tag{19}$$

where γ_b is the total shear strain up to fracture expressed in engineering units. In the continuous chain model for the fibre extension the tensor notation is used, which means that $\gamma = 2(\Theta - \theta)$. Using Hooke's relation $\tau = g\gamma$, the expression for the work or the shear strain energy up to fracture becomes

$$W_b^S = 2\int_0^{\theta_b} \tau d(\Theta - \theta) = 4g\int_0^{\theta_b} (\Theta - \theta)\, d(\Theta - \theta) = 2g(\Theta - \theta_b)^2 \tag{20}$$

where θ_b is the angle of the chain axis at fracture. To a first approximation the shear stress at fibre fracture is given by $\tau_b = 2g(\Theta - \theta_b)$, so the work of fracture can also be written as

$$W_b^S = 2g(\Theta - \theta_b)^2 = \frac{\tau_b^2}{2g} \tag{21}$$

By neglecting the chain extension it is assumed that $W_b \approx W_b^S$. Thus the observation that W_b is approximately constant for fibres of the same polymer with different degrees of orientation means that not only a constant critical shear stress, τ_b, but also a maximum shear strain, $(\Theta - \theta_b)$, is a useful criterion of fibre failure, irrespective of the modulus of the fibre. The fracture envelope $\{\varepsilon_b, \sigma_b\}$ is now calculated on the basis of the continuous chain model. As a first approximation the contribution from the chain extension is neglected and the critical shear stress criterion will be employed. According to the model the shear stress at fracture is given by

$$\tau_b = -\sigma_b \sin\theta_b \cos\theta_b \tag{22}$$

and the shear strain by

$$\tan(\theta_b - \Theta) = -\frac{\sigma_b}{2g} \sin\theta_b \cos\theta_b \tag{23}$$

where σ_b is the stress at fracture or the strength of the filament and $\beta=(\Theta-\theta_b)$ is the fracture shear strain. The fibre strain at fracture due to the shear deformation, ε_b^s, is given by the change in the length of the chain segment projected onto the filament axis

$$\varepsilon_b^s = \frac{\cos\theta_b}{\cos\Theta} - 1 \tag{24}$$

The relation between the end points of the tensile curve, σ_b and ε_b $(=\varepsilon_b^s)$, can be calculated with Eqs. 9, 23 and 24. This relation is now by definition taken as the fracture envelope. Note that these equations only hold for elastic deformation. In order to account for some viscoelastic and plastic deformation, a value g_v is used, which is somewhat smaller than the value for elastic deformation g. The dashed curves in Figs. 8 and 9 are the calculated fracture envelopes (neglecting the chain extension) for the cellulose II and the POK fibres, respectively. These figures show a good agreement between the observed and calculated fracture points.

The concept of a maximum shear strain is supported by the experimental relationship for the lifetime of a polymer fibre. For many polymer fibres the observed lifetime or the time to failure t_b is given by

$$\log(t_b) = C_1 - C_2\sigma \tag{25}$$

where σ is the creep stress and the parameters C_1 and C_2 are a function of temperature. For example, a PpPTA fibre breaks at room temperature after 10^4 s at a creep stress of 3 GPa, after 10^3 s at a creep stress of 3.17 GPa and after 10^2 s at a stress of 3.31 GPa [30]. In fact, Eq. 25 expresses that the strength is a function of time. This relation cannot be explained by assuming a constant critical shear stress as a criterion of fibre fracture, because in a creep failure experiment the applied stress is lower than the fibre strength determined in a normal tensile test.

As shown in Fig. 10, the shear strain is proportional to the relative displacement of two parallel aligned adjacent chains. Therefore, it seems plausible to assume that the maximum shear strain value $\beta=(\Theta-\theta_b)$ at which fracture of the fibre is initiated will be related to a critical overlap length between adjacent

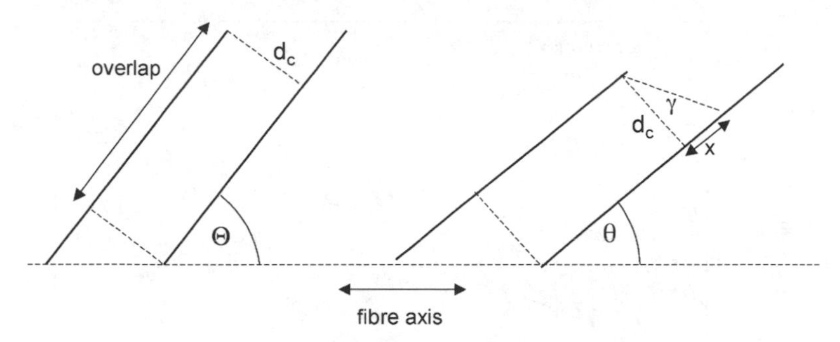

Fig. 10 Shear deformation of a domain shown in Figs. 5 and 6 results in a rotation of the chain axes and in a relative displacement x of the chains, which is proportional to the shear angle or the shear strain $\gamma=2(\Theta-\theta)$

polymer chains. Part of the maximum or critical shear strain is brought about by the elastic shear deformation, and the remaining part is due to creep shear deformation. In the case of a normal tensile test using relatively high strain rates (100% min^{-1}) for the determination of the fibre strength, the duration of the test is short and the creep shear contribution is very small indeed. Consequently, in this test the interchain bonds are broken within a very short time. In this particular case Eq. 21 shows that the criterion for fibre failure at a critical shear strain corresponds to failure at a critical shear stress.

In the case of fibres made of flexible-chain polymers, e.g. polyamide 6 and 66 and poly(*p*-ethylene terephthalate) or PET, the fracture criterion involving a critical shear strain cannot be employed without modifications. These fibres have a two-phase structure consisting of a series arrangement of amorphous and crystalline domains. When for increasing fibre stress the critical shear stress is reached – presumably at the (second) maximum of the modulus–strain curve of the fibre – the chains in the amorphous domains start to flow and unfolding of the chains occurs. It has been shown that, up to this maximum, the continuous chain model can accurately describe the tensile curve of polymer fibres [1–10]. For decreasing modulus the fracture envelope of these fibres progressively deviates from the hyperbolic shape due to the contribution of plastic flow to the fibre strain, as is shown for PET fibres in Fig. 11. For medium- and low-oriented fibres made of flexible-chain polymers the viscoelastic and plastic rotation of the chain together with the unfolding of chains form the major contribution to the fibre extension.

A simple approximation for the strength of a fibre is derived by exclusion of the chain extension. By using the analytical function Eq. 9 as the approximation of the solution of Eq. 23

$$\tan\theta_b = \frac{\tan\Theta}{\left(1+\dfrac{\sigma_b}{2g}\right)} \qquad (26)$$

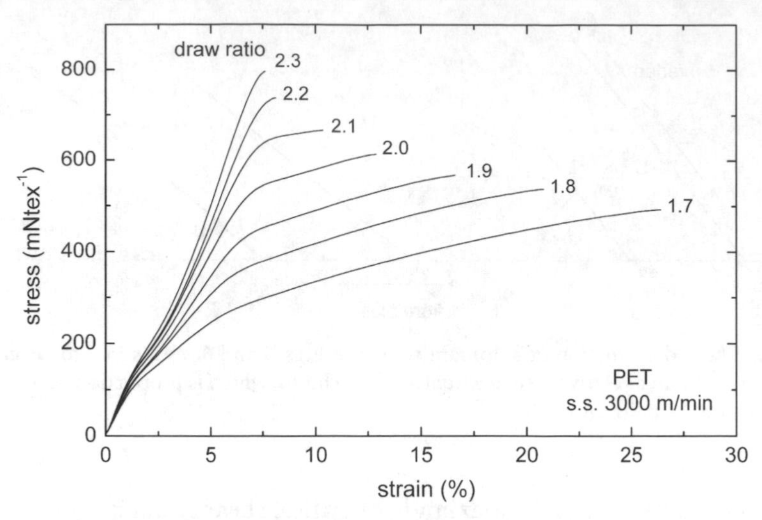

Fig. 11 Tensile curves of PET yarns made with different draw ratios

the fracture condition becomes

$$|\tau_b| = \sigma_b \sin\theta_b \cos\theta_b \approx \frac{\sigma_b \tan\Theta}{\left(1 + \dfrac{\sigma_b}{2g}\right)} = 2g\beta \tag{27}$$

which yields for the fibre strength, σ_b^s, neglecting the contribution from chain extension

$$\sigma_b^s = \frac{2g\beta}{\tan\Theta - \beta} \tag{28}$$

This equation, which only holds for initial orientation angles $\Theta > \arctan\beta$, shows that the fibre strength increases with decreasing orientation angle of the chains and that it is proportional to g and thus proportional to the critical shear stress τ_b. Equation 28 is in agreement with the observation by Knoff. He found a linear relation between the tensile strength of a PpPTA filament and its torsional shear strength, indicating the importance of shear failure for the tensile strength of a PpPTA fibre [31].

This simple fracture model has a major shortcoming. The exclusion of chain stretching in the model leads for small initial orientation angles to strength values that become infinite. It follows from Eq. 27 that the shear stress is a continuous function of the fibre stress and it increases asymptotically to the value of $2g\tan\Theta$. So for initial orientation angles

$$\Theta < \arctan\beta = \arctan\left(\frac{\tau_b}{2g}\right) \tag{29}$$

the shear stress acting on a domain will never reach the critical value τ_b. Consequently the fibre strength in this simple fracture model becomes infinite. Thus a model based on a critical shear stress alone, without taking account of the chain extension, cannot provide the complete description of the tensile strength of polymer fibres.

However, considering the simplicity of the model using a single orientation angle, the agreement between the observed and computed fracture envelopes is still surprisingly good. It shows that for well-oriented yarns a higher fibre stress is required to exceed the critical shear stress τ_b ($|\tau|=\sigma\sin\theta\cos\theta\geq\tau_b$) than for low-oriented fibres. Moreover, as the shear stress increases with the orientation angle, the onset of fracture is likely to be in the domains with angles in the tail of the orientation distribution. The basic assumption of the model is that fracture of fibres is due to the rupture of the intermolecular bonds as a result of the shear deformation. This is confirmed by the observation that the fracture morphology of polymer fibres that do not have a melting temperature has a more or less fibrillar nature, i.e. they are broken by shear fracture.

2.3 Distribution of the Strain Energy in a Fibre

Owing to the chain orientation distribution the fracture mechanism of oriented polymer fibres is different from that of isotropic fibres. The presence of this distribution leads to a non-uniform distribution of the strain energy between the domains. The strain energy is defined by

$$W=\int_0^{\varepsilon}\sigma d\varepsilon=\int_0^{\sigma}\frac{\sigma d\sigma}{E} \tag{30}$$

For a polymer fibre with a single orientation angle the modulus, E, or the slope at each point of the tensile curve, is a function of the tensile stress and given by

$$\frac{1}{E(\theta)}=\frac{1}{e_c}+\frac{\sin^2\theta}{(2g+\sigma)} \tag{31}$$

where θ is the angle of the chain axis at a fibre stress σ [4–6]. For well-oriented fibres the approximation $\sin^2\theta\approx\tan^2\theta$ is used, and the integral in Eq. 30 can be evaluated with Eq. 9

$$W\approx\frac{\sigma^2}{2e_c}+(g\tan^2\Theta)\cdot\left(\frac{\sigma}{2g+\sigma}\right)^2 \tag{32}$$

Equation 32 gives the total strain energy stored in a domain of a fibre with an orientation angle Θ in the unloaded state after the stress has been increased from 0 to σ. The first term on the right-hand side is the strain energy of the chain extension, and the second term is the shear strain energy. The continuous chain model postulates that a fibre can be considered as a series arrangement of domains whose symmetry axes are distributed in accordance with the orientation distribution of the chains. As shown by Eq. 32 the strain energy per domain during tensile deformation is not uniformly distributed over the whole fibre, but is proportional to the function $\tan^2\Theta$. If $\rho(\theta)$ is the distribution measured along the meridian in the unloaded state, the total number of domains in the fibre at an angle Θ with the fibre axis is proportional to the function $N(\Theta)=\rho(\theta)\sin\Theta$ [1]. Figure 12 demonstrates that for a Gaussian distribution $\rho(\theta)$ with a width at half-height of 17.2°, the most frequent orientation angle is near 7.5°. The distribution of the strain energy between all domains in the fibre is given by the function $W(\Theta)=\tan^2\Theta\sin\Theta\rho(\Theta)$, which for this Gaussian distribution shows a maximum at about 12.5°. The angle of $\Theta=12.5°$ is in the tail of the Gaussian distribution. The ratio of the strain energies of domains with orientation angles at $\Theta=12.5°$ and 5° equals $(\tan^2 12.5)/\tan^2 5=6.4$. As shown in Fig. 12, for a Gaussian distribution with a width at half-height of 17.2° the ratio of the strain energies $W(\Theta)$ stored in all domains at these angles is equal to 5. This example demonstrates that the major part of the strain energy supplied in a tensile test to the fibre is stored in the tail of the distribution $\rho(\Theta)$, which has important implications for the initiation of fracture.

Immediately upon fracture the fibre drops from a high-energy state equal to the stored elastic energy to its lowest energy, viz. the unloaded state. Hence, initiation of fracture in the domains in the tail of the orientation distribution $\rho(\theta)$ does release most effectively the stored energy of a loaded polymer fibre. So, if there are no impurities and structural irregularities, fracture of the fibre is

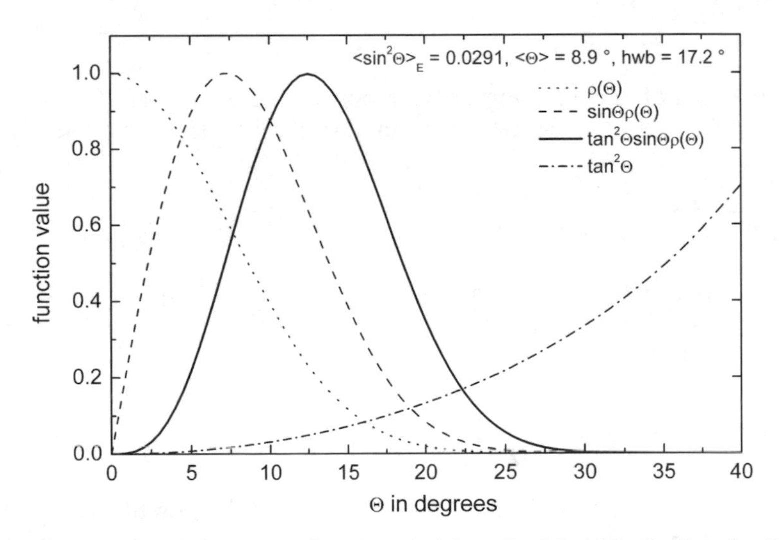

Fig. 12 The meridianal orientation function $\rho(\Theta)$ for a Gaussian distribution, the distribution function $\rho(\Theta)\sin\Theta$, the shear strain energy function $\tan^2\Theta\sin\Theta\rho(\Theta)$ and the function $\tan^2\Theta$

likely to begin in domains lying in the tail of the orientation distribution. This means that upon fracture the secondary bonds in the domains with smaller orientation angles are not strained to the critical shear angle. In Sect. 2.1 we arrived at the conclusion that in the case of the cellulose II fibres only about 7% of the hydrogen bonds need to be broken for fracture of the fibre. Apparently this relatively small fraction originates from the non-uniform distribution of the strain energy in the fibre. Presumably most of the broken hydrogen bonds are located in the domains with a large orientation angle.

In the previous section we discussed the ultimate strength of a polymer fibre σ_0. This value corresponds to the stress at which all secondary bonds in the fibre are broken. Due to the presence of the chain orientation distribution alone, it follows that even the strength of a polymer fibre without any flaws will never attain this value. Yet, fracture in a real fibre may not always initiate in the most disoriented domains. If there are inhomogeneities that lead to stress concentrations, fracture can also occur in domains at a smaller angle to the fibre axis.

2.4 Effect of Cracks on the Strength and the Relation with the Chain Orientation

Organic polymer fibres contain impurities and small cracks or voids. In particular, in wet-spun fibres these voids are oriented more or less parallel to the chain direction, as has been revealed by small-angle X-ray scattering. Presumably their number and size can influence the critical shear stress at which the fibre will break. In general, small surface cracks in these fibres that are perpendicular to the fibre axis have little effect on the strength. As in uniaxially oriented macrocomposites, a small crack in the surface transverse to the filament axis will bend off in a direction parallel to the chain direction due to the Cook–Gordon mechanism [32]. At the front of the crack tip in these anisotropic materials there is not only a concentration of the tensile stress but also a concentration of the shear stress parallel to the direction with the highest modulus, which is usually the fibre axis. Due to this shear stress elongated voids parallel to the fibre axis are created at the front of the transverse crack tip. Such a void may coalesce with the crack tip, thereby substantially increasing the curvature of the crack tip and thus lowering the stress concentration.

For an introduction to the subject of the strength of materials the reader is referred to the book "Structures, or why things don't fall down" by J.E. Gordon [33]. He states that in order to break a material in tension a crack must spread right across it. However, creating a new crack requires a supply of energy. This energy can be tapped from the stored-up strain energy (or resilience) due to the stress on the material. According to the modern view on strength, when we break a structure or a material by loading in tension, we ought not to regard fracture as being caused directly by the action of the applied load pulling on the chemical bonds between the atoms in the material. That is to say, it is not the consequence of the simple action of a tensile stress as the classical textbooks would make us believe. The "true" or theoretical maximum tensile stress required to pull the atoms apart is very high indeed, far higher than the "practical" strength determined by means of ordinary tensile tests. The direct result of increasing the load on a structure or a material is only to cause more strain energy to be stored within it. The significant question related to the strength of a material is whether or not it is possible for its strain energy to be converted into fracture energy so as to create a new crack. Thus, the tensile fracture of a polymer fibre depends chiefly upon: (1) the price in terms of energy which has to be paid in order to create a new crack, (2) the amount of strain energy which is likely to become available to pay this price and (3) the size and shape of the worst hole or crack in the structure or material.

According to the theory of crack growth by Griffith the balance between the stored strain energy and the work of fracture necessary to create the fracture surfaces controls the propagation of cracks [18, 19]. The energy required to propagate the crack is delivered by the release of the strain energy in the areas around the crack. Figure 13 shows an isotropic piece of material under stress with a small crack normal to the stress direction. After a crack of length L is formed, the strain energy in roughly the shaded area around the crack has been released. This strain energy, which is proportional to L^2, is consumed as work of fracture to create the crack. This work of fracture is proportional to L [33]. From the balance between the strain energy and the work of fracture it follows that the so-called Griffith length can be defined

$$L_G = \frac{w_\gamma E}{\pi\sigma^2} \tag{33}$$

where w_γ is the work of fracture or the surface energy of the crack per unit volume, E the modulus and σ the applied stress. For cracks with a length $L<L_G$ the system is energy-consuming and cracks will not propagate. For an infinitesimal increase of the crack, the increase of the work of fracture is larger than the increase of the strain energy which is needed to deliver the energy for the fracture. Cracks larger than L_G are self-propagating and will eventually result in fracture of the material. Equation 33 shows that the critical crack length or the Griffith length is a function of the modulus and the stress on the material.

Although Eq. 33 has been derived for isotropic materials, it will be applied here as a first approximation to polymer fibres. For a fibre subjected to an ax-

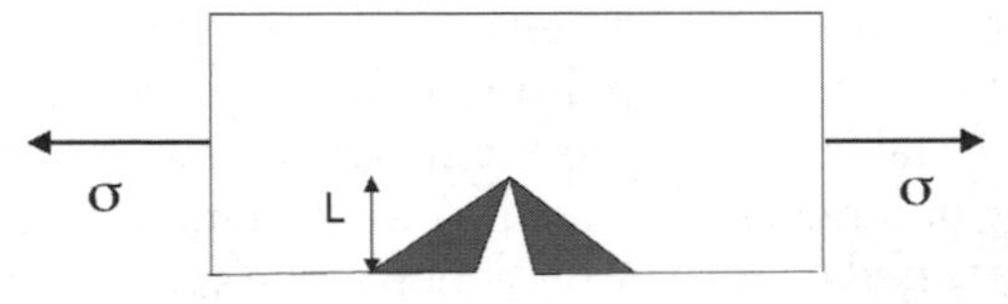

Fig. 13 The strain energy of the *shaded area* has been consumed as work of fracture for the formation of a crack with length L

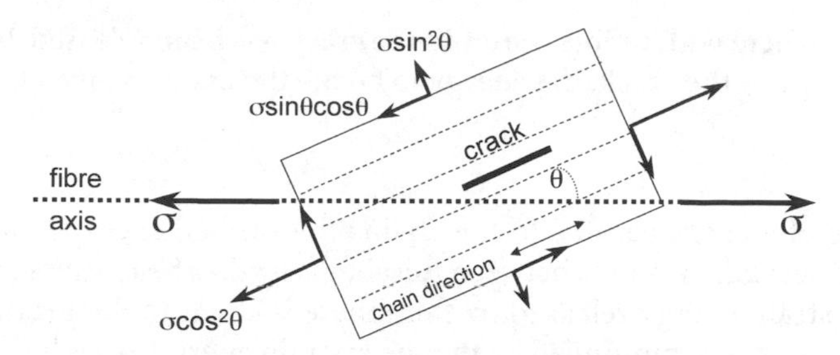

Fig. 14 The four normal stresses and the four shear stresses acting on a domain resulting from a tensile stress on the fibre. A crack is drawn parallel to the chain direction. The angle in the loaded state between the domain axis and the fibre axis is θ

ial tensile stress Fig. 14 shows the stresses acting on a domain with a crack parallel to the chain direction. There are two equal and opposing stresses, $\sigma\cos^2\theta$, parallel to the chain direction and two equal and opposing stresses, $\sigma\sin^2\theta$, normal to the chain direction. Furthermore, the domain is subjected to four shear stresses, $\tau=-\sigma\sin\theta\cos\theta$, causing a shear deformation of the domain parallel to the chain direction. Hence two Griffith lengths for cracks parallel to the chain axis can now be defined in a fibre, viz. one caused by the stress transverse to the chain direction and one caused by the shear stresses acting on the domain.

First an expression of the Griffith length for the shear stresses is derived by taking into account the shear strain energy of the domain. An elongated crack is oriented parallel to the chains. Suppose the crack has a circular shape with a radius q as shown in Fig. 15. Due to the formation of the crack the strain

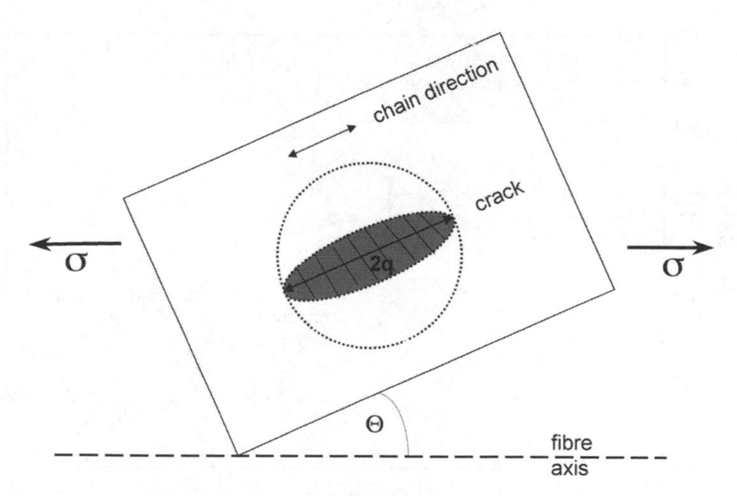

Fig. 15 Initiation of fibre fracture by a crack oriented parallel to the chain direction in a domain. It is proposed that a circular crack with a radius q releases the strain energy in a sphere around the crack with the same radius. Note that in this two-dimensional drawing only the circular crack is shown in perspective

energy in a sphere with radius q around the crack has been released. If w_γ^S is the surface energy of the crack, the energy to create the crack is given by

$$W_1 = 2\pi q^2 w_\gamma^S \tag{34}$$

In the case of a crack parallel to the chains the surface energy relates to the cleavage of secondary bonds between the chains by the shear stress $\sigma\sin\theta\cos\theta$. The shear strain energy released by the sphere is equal to the product of the volume of the sphere multiplied by the shear strain energy given by the second term in Eq. 32. Thus

$$W_2 = \frac{4\pi q^3 \sigma^2 g \tan^2\Theta}{3(2g+\sigma)^2} \tag{35}$$

In fact at least this amount of strain energy is needed for the formation of the circular crack with radius q. The total free energy contribution due to the presence of the crack is

$$\Delta W = W_1 - W_2 \tag{36}$$

As shown in Fig. 16 the condition that the crack spreads is

$$\frac{\partial \Delta W}{\partial q} = 0 \tag{37}$$

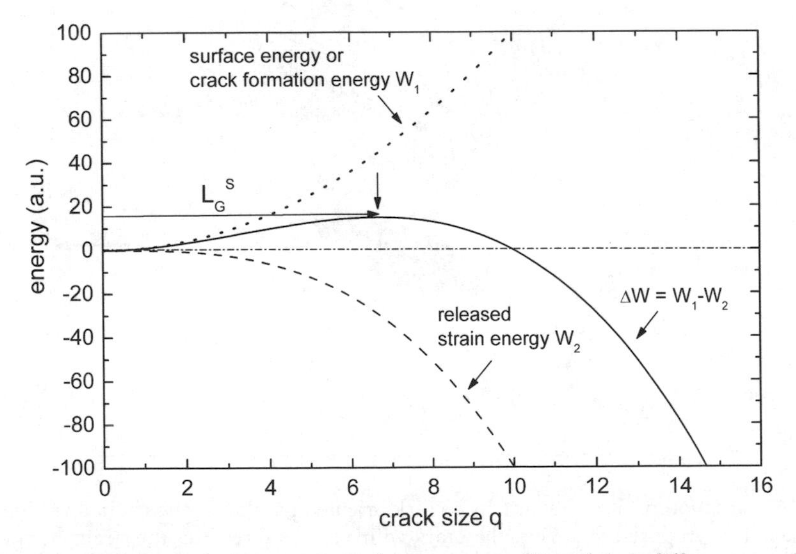

Fig. 16 Graph of the strain energy W_2, the fracture energy W_1 and their difference, resulting in the Griffith length L_G^S. Cracks with sizes larger than L_G^S will propagate in the fibre

which gives for the largest size of a crack that is not self-propagating

$$L_G^S = \frac{w_\gamma^S}{\tan^2\Theta}\left(\frac{4g}{\sigma^2} + \frac{4}{\sigma} + \frac{1}{g}\right) \tag{38}$$

This equation shows that for increasing angle Θ the Griffith length, L_G^S, decreases.

In a similar way the Griffith length, L_G^T, due to the strain energy caused by the stress transverse to the chain direction can be derived

$$L_G^T \approx \frac{2w_\gamma^T e_1}{3\tan^4\Theta}\left(\frac{1}{\sigma^2} + \frac{2}{g\sigma} + \frac{3}{2g^2}\right) \tag{39}$$

In this expression e_1 is the modulus perpendicular to the chain axis and w_γ^T is the work of fracture for the transverse tensile stress. Equation 39 demonstrates that the Griffith length L_G^T increases rapidly with decreasing orientation of the chain angle Θ.

Comparison of Eq. 38 with Eq. 39 indicates that, due to the difference between the angle-dependent factors, L_G^T is likely to be considerably larger than L_G^S. Thus, for cracks parallel to the chain axis the Griffith length related to the shear stress is more relevant to the fracture of polymer fibres than the Griffith length related to the normal stress $\sigma\sin^2\theta$. Furthermore it shows that a polymer fibre with a broad orientation distribution is more prone to failure by cracks than a fibre with a contracted distribution. This can explain the observation made by Penning et al. mentioned in Sect. 1, that longitudinal size effects in high-strength UHMW PE fibres become weaker as the modulus increases.

In conclusion, the initiation of fracture in a polymer fibre preferably occurs in the domains in the tail of the orientation distribution. The reasons are: (1) in these domains the local shear stress will exceed the critical shear stress first, (2) the release of the strain energy is most effectively brought about by fracture of these domains and (3) the Griffith length in these domains adopts its lowest value.

2.5 The Polymer Fibre Considered as a Molecular Composite

The simple model presented in Sect. 2.2 results in an infinite strength for orientation angles smaller than $\arctan(\tau_b/2g)$. An improved theory for the strength of polymer fibres is provided by the molecular composite approach as has been suggested by Knoff [31]. He observed that the tensile fracture morphology of aramid fibres is often highly fibrillated and analogous to that of a uniaxially oriented fibre-reinforced composite that fails in tension via matrix shear failure initiated at the fibre ends. Therefore the fibre will now be regarded as a molecular composite of chains embedded in a matrix of secondary interchain bonds. An expression for the fibre strength is derived from the expression for the strength of a macrocomposite composed of parallel oriented and continuous filaments in a matrix. Based on one of the best-known failure criteria, viz. the maximum work theory, also commonly referred to as the Tsai–Hill criterion, the strength of a macrocomposite σ_{comp} is given by

$$\frac{1}{\sigma_{comp}^2} = \frac{\cos^4\theta}{\sigma_L^2} + \left(\frac{1}{\tau_b^2} - \frac{1}{\sigma_L^2}\right)\sin^2\theta\cos^2\theta + \frac{\sin^4\theta}{\sigma_T^2} \tag{40}$$

where σ_L is the longitudinal strength of the filaments, τ_b the shear strength of the filament/matrix interface, σ_T the strength transverse to the filament direction and θ the orientation angle of the filament relative to the stress direction [34]. The strength values σ_L, σ_T and τ_b are considered to be the limiting stresses of linear elastic behaviour. Equation 40 takes account of the interactions between the three failure stresses. In the macrocomposite model it is assumed that the filaments are continuous, i.e. σ_L is the strength of the filament of which the macrocomposite is composed.

The polymer fibre is considered to be composed of a parallel array of identical fibrils. A single fibril is a series arrangement of domains with a varying angle to the fibre axis described by the orientation function $\rho(\Theta)$ in the unloaded state. The strength of a domain is supposed to be given by

$$\frac{1}{\sigma_b^2} = \frac{\cos^4\theta_b}{\sigma_L^2} + \left(\frac{1}{\tau_b^2} - \frac{1}{\sigma_L^2}\right)\sin^2\theta_b\cos^2\theta_b + \frac{\sin^4\theta_b}{\sigma_T^2} \tag{41}$$

where θ_b is the angle of the chain axis at fracture given by Eq. 26, σ_L the strength of the domain along the chain direction, τ_b the shear strength of interchain bonding and σ_T the strength transverse to the chain axis. Thus, at a stress σ_b the domains with an orientation angle θ_b will start to crack. When these cracks coalesce with cracks in neighbouring fibrils the local stress on the domains with smaller orientation angles increases. Subsequently they will crack and as a result the failure process in the fibre is initiated.

In analogy with a uniaxially oriented macrocomposite, it is now postulated that the strength of the fibre can be approximated by the expression

$$\frac{1}{\sigma_b^2} \approx \frac{\langle\cos^4\theta_b\rangle}{\sigma_L^2} + \left(\frac{1}{\tau_b^2} - \frac{1}{\sigma_L^2}\right)\langle\sin^2\theta_b\cos^2\theta_b\rangle + \frac{\langle\sin^4\theta_b\rangle}{\sigma_T^2} \tag{42}$$

where the average is taken over the orientation distribution at fracture $\rho(\theta_b)$. For well-oriented fibres the term $(\sin^4\theta)/\sigma_T{}^2$ may be neglected. For a Gaussian orientation distribution the following approximations can be made

$$\begin{aligned}
&\langle\cos^4\theta_b\rangle \approx 1 - 2\langle\sin^2\theta_b\rangle + 2(\langle\sin^2\theta_b\rangle)^2 \\
&\langle\sin^2\theta_b\cos^2\theta_b\rangle \approx \langle\sin^2\theta_b\rangle - 2(\langle\sin^2\theta_b\rangle)^2 \\
&\langle\sin^4\theta_b\rangle \approx 2(\langle\sin^2\theta_b\rangle)^2
\end{aligned} \tag{43}$$

In order to avoid intricate computations involving the calculation of the orientation distribution at fracture, it is assumed that the orientation distribution of the chains can be characterised by a single angle. This assumption excludes the effects of the orientation distribution on the fibre strength as discussed in the previous section. Yet, we believe that for the derivation of general trends this simplification is allowed. By applying Eqs. 26, 42 and 43 the strength can be calculated as a function of the orientation parameter in the unloaded state, and thus as a function of the initial modulus.

2.5.1
Ultimate Strength of a Fibre

As shown by Yoon the ultimate strength σ_L of a fibre composed of chains of finite length, which are perfectly oriented along the fibre axis, is determined by the strength of the intermolecular bonding [11]. Due to the finite length a shear stress arises at the chain end which, upon exceeding a critical stress, causes debonding of the chain from its nearest neighbours. This occurs at a stress that is much lower than the breaking strength of the chain itself. Hence the mechanical load transfer between polymer chains is through intermolecular interaction, which acts similar to that of a shear stress, and the fibre strength is primarily governed by the intermolecular adhesion strength. Therefore, as shown in Fig. 17, σ_L should be some function of the shear strength τ_0 and of the chain length distribution.

As will be shown in Sect. 3, Yoon derived the following expression for the ultimate strength of a polymer fibre with long and parallel oriented chains of finite length

$$\sigma_L = \frac{2V_c\tau_0}{\mu} \quad \text{with} \quad \mu^2 = \frac{3.1 \cdot g}{e_c} \tag{44}$$

or

$$\sigma_L = 1.14 \cdot \tau_0 \sqrt{\frac{e_c}{g}} \tag{45}$$

where τ_0 is the ultimate shear strength at which the chain separates from the matrix and the separated interface does not support any load, V_c is the chain

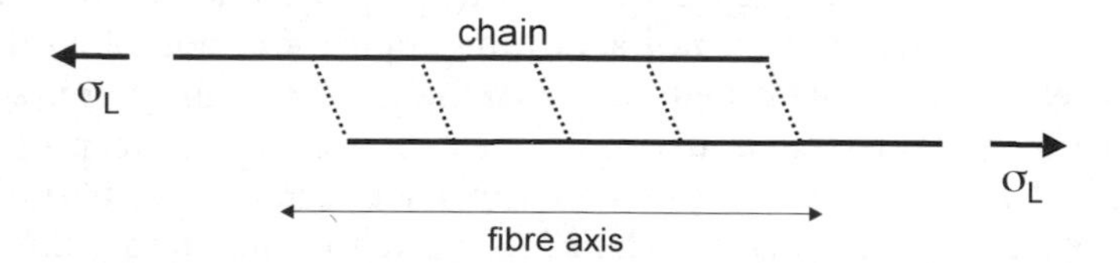

Fig. 17 The ultimate strength of a fibre with perfectly parallel oriented chains of finite length is determined by the strength of the secondary bonding

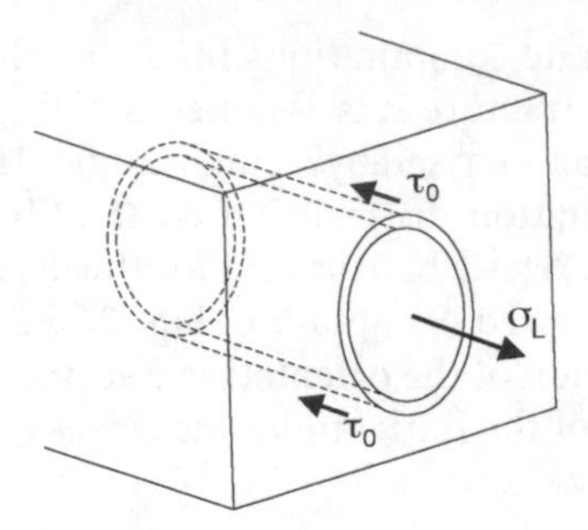

Fig. 18 The strength σ_L and the ultimate shear strength τ_0 for debonding of the chain from the surrounding matrix consisting of secondary bonds

volume fraction of the fibre and equal to 1, g is the modulus for shear between the chains and e_c is the chain modulus [11, 35]. As shown in Fig. 18, at a stress σ_L the chain is pulled out from the surrounding matrix consisting of secondary bonds.

Yoon assumed a constant shear modulus, $\tau_0=g\gamma_b=2g\beta$; thus the corresponding fracture energy is, according to Eq. 21, $W_0^S=\tau_0^2/(2g)$. This results in the following value of the ultimate strength

$$\sigma_L = 2.3 \cdot \beta\sqrt{ge_c} \tag{46}$$

or in terms of the shear energy at fracture

$$\sigma_L = 1.6 \cdot \sqrt{e_c W_b^S} \tag{47}$$

The estimate of the theoretical strength of a fibre based on the debonding strength of the intermolecular bonds between the chains is given by Eqs. 45 and 46. It differs fundamentally from the estimate based on the breaking strength of a single chain. The experimental value of the breaking strength of a Si–C covalent single bond is about 12 GPa [36]. Based on the enthalpy of dissociation, the C–C and C–N bonds are of the same strength. As an example, the PpPTA fibre is chosen with e_c=240 and g=2 GPa. A reasonable estimate of the critical shear stress is 20% of g or τ_0=0.4 GPa, which yields σ_L=5 GPa. The present strength of a PpPTA yarn with a modulus of 100 GPa is about 4 GPa. Comparison of these values with the breaking strength of a single chain strongly suggests that shear failure initiates the fracture of the PpPTA fibre, although chain fracture due to local stress concentration around impurities cannot be ruled out.

For an estimate of the ultimate shear strength, τ_0, of a single domain based on the lattice parameters we use a simple shear plane system proposed by Frenkel [19]. As shown in Fig. 19 it consists of a linear array of periodic force centres resembling the polymer chain. According to this model the relation between the relative displacement x along the shear direction and the shear stress is given by

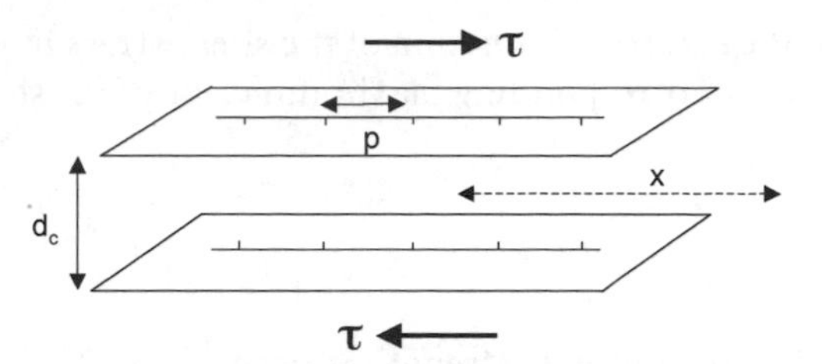

Fig. 19 Shear plane system with periodic force centres spaced at a distance p along the shear direction x and with an interplanar spacing d_c according to the model of Frenkel [19]

$$\tau(x) = \tau_m \sin\left(2\pi \frac{x}{p}\right) \quad 0 \le x \le p/4 \tag{48}$$

where p is the periodic distance between the force centres along the shear direction. Thus

$$\partial\tau = \frac{2\pi\tau_m}{p} \cos\left(2\pi \frac{x}{p}\right) \partial x \tag{49}$$

For an interplanar spacing, d_c, the shear angle in engineering notation is given by $\partial\gamma = \partial x / d_c$. As the shear modulus is defined by $g = \partial\tau/\partial\gamma$, it follows that

$$g(x) = \frac{2\pi d_c \tau_m}{p} \cos\left(2\pi \frac{x}{p}\right) \tag{50}$$

Hence, the shear modulus for small values of x is

$$g = \frac{2\pi d_c \tau_m}{p} \tag{51}$$

Note that in this model $g=0$ for $x=p/4$. The shear energy, W^S, for a shear displacement x is given by

$$W^S(x) = \int_0^\gamma \tau d\gamma = \frac{pg}{2\pi d_c} \int_0^x \sin\left(2\pi \frac{x}{p}\right) \frac{\partial x}{d_c} \tag{52}$$

or

$$W^S(x) = \frac{p^2 g}{4\pi^2 d_c^2} \left[1 - \cos\left(2\pi \frac{x}{p}\right)\right] \quad 0 \le x \le p/4 \tag{53}$$

The functions $\tau(x)$ and $W^S(x)$ are depicted in Fig. 20. The maximum shear stress τ_m, which occurs for $x=p/4$ or at a (maximum) shear strain of $p/(4d_c)$ is defined as the ultimate shear strength τ_0

$$\tau_0 = \frac{p}{2\pi d_c} g \tag{54}$$

(Note that due to the sinusoidal function of the shear stress in Frenkel's model $\tau_m \neq g\gamma_b = gp/(4d_c)$). The corresponding maximum value of the shear energy, W_m^S, is given by

$$W_m^S = \frac{p^2 g}{4\pi^2 d_c^2} \tag{55}$$

From Eqs. 45 and 55 the ultimate strength is derived

$$\sigma_L = \frac{0.57 \cdot p}{\pi d_c} \sqrt{g e_c} \tag{56}$$

or in terms of the strain energy

$$\sigma_L = 1.14 \cdot \sqrt{e_c W_b^S} \tag{57}$$

The difference between Eqs. 47 and 57 results from the difference between $W_0^S = \tau_0^2/(2g)$ of Yoon's model and $W_m^S = \tau_m^2/g$ according to Frenkel's model. In terms of the critical shear strain $\gamma_b = 2\beta$ the difference between Eqs. 46 and 56 is small. As has been shown in Sect. 2.3, during tensile deformation of a real fibre the strain energy supplied by the tensile tester is mainly channelled into the domains with the largest orientation angles. Therefore initiation of fracture is localised in these domains. Fracture of these domains creates a crack, which brings about an increase of the stress in the remaining part of the fibre, thereby raising the fraction of domains in which the fracture condition is reached. So

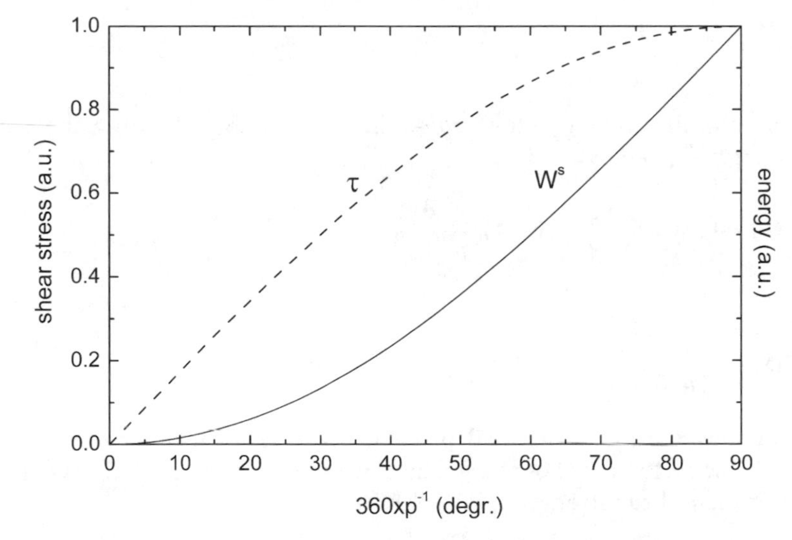

Fig. 20 Shear stress τ and shear energy W^S as a function of shear displacement x for a simple shear plane system

an acceleration of the crack-forming process takes place and fracture of the whole fibre is initiated. Hence, the strain energy per unit mass W_b required to fracture a real polymer fibre will always be smaller than the maximum debonding energy, W_0^S or W_m^S, calculated for the single domain. In fact, the calculation of the theoretical strength based on the debonding energy, W_0^S or W_m^S, yields a value which refers to the complete separation of all chains in a perfectly ordered domain free of any impurities.

Impurities and flaws have a detrimental effect on the fibre strength. Due to shear stress concentrations at structural irregularities and impurities, the ultimate debonding stress τ_0 ($\sim\tau_m$) or the critical fracture strain β may be exceeded locally far sooner than in perfectly ordered domains. Thus, during the fracture process of real fibres the debonding from neighbouring chains occurs preferably in the most disoriented domains and presumably near impurities. At the same time, however, the chains in the rest of the fibre are kept under strain but remain bonded together up to fracture.

Equation 54 shows that a large shear modulus does not necessarily mean a large debonding stress. Depending on the ratio p/d_c it is very well possible in the Frenkel model to have a small shear modulus in conjunction with a large debonding stress. But when the dimensions p and d_c are kept constant, Eq. 54 shows that the shear strength τ_0 is directly proportional to g. In the case of metals and inorganic solids the estimated ratio $\gamma_b = \tau_0/g$ varies from 0.1 to 0.25 [19].

So far we have employed in this discussion a critical shear stress as a criterion for fibre fracture. In Sect. 4 it will be shown that a critical shear strain or a maximum rotation of the chain axis is a more appropriate criterion when the time dependence of the strength is considered.

The Frenkel model indicates that the shear planes play a crucial role in the fracture process. The information on the most likely slippage or shear planes is provided by the crystal structure. In Sect. 2.5.2 the structures of cellulose II, PpPTA and PIPD-HT (M5-HT) fibres are considered. The periodicity of the function describing the shear stress in the Frenkel model is the result of a crystalline perfection that extends over large distances compared to the dimensions of p and d_c. This regularity of the bonding between the shear planes reinforces the bonding between neighbouring chains, which can be described as a co-operative effect. Originally the Frenkel model was derived for metals and ionic solids. In these materials the dimensions of p and d_c are in fact the distances between neighbouring atoms. Therefore, they are several orders of magnitude smaller than the size of the crystallites, which are usually larger than 1 μm. Thus, in these materials the crystalline order is almost perfect compared to that in the polymer fibres. The intrinsic imperfections in polymer fibres are due to the chain length distribution and the flexibility of the chain. Also irregularities in the chain conformation contribute to structural disorder. Therefore, the periodicity of the sinusoidal function describing the shear stress in the Frenkel model is likely to be disturbed to some degree by, for example, the chain ends shown in Fig. 21.

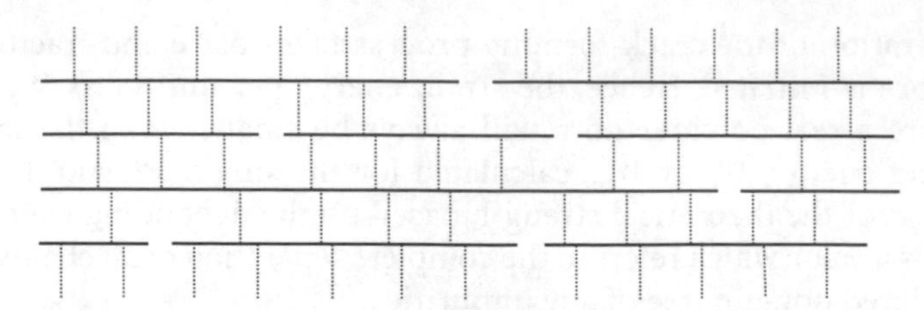

Fig. 21 Schematic visualisation of interchain bonding in a domain of a polymer fibre. The *lines* represent the chains with a length distribution and the *dashed lines* depict the interchain secondary bonds. Chain ends prevent the formation of these bonds in their immediate surroundings

This raises the question how much the disorder will diminish the co-operative effect and therefore the strength of the interchain bonding. In other words, how much is the critical shear stress influenced by the chain length distribution? Therefore, the application of the proposed equations for polymer fibres requires some caution. The Frenkel model does not take into account the chain length of the polymer. Presumably there is a relation between critical shear strength and average chain length in the fibre. As there is no information on the intermolecular or cohesive energy of polymer fibres and its dependence on the average chain length, it is proposed to use the activation energy, W_a, for creep in polymer fibres as an approximation of the maximum shear energy $W_m{}^S$. This can be inferred from the observation of Wu et al. that the strength values of PpPTA fibres obtained from a tensile test were considerably lower than the values obtained from extrapolation of creep failure experiments [30]. Part of the deformation process in a rapid tensile test up to fracture involves the (early) breakage of secondary bonds due to the stress concentrations around flaws and impurities localised in the domains in the tail of the orientation distribution. In a creep experiment, however, relaxation of the stress concentrations at structural inhomogeneities can take place. This might be considered as if a tensile test is performed on a more perfect fibre.

Estimates of the ultimate shear strength τ_0 can be obtained from molecular mechanics calculations that are applied to perfect polymer crystals, employing accurate force fields for the secondary bonds between the chains. When the crystal structure of the polymer is known, the increase in the energy can be calculated as a function of the shear displacement of a chain. The derivative of this function is the attracting force between the chains. Its maximum value represents the breaking force, and the corresponding displacement allows the calculation of the maximum allowable shear strain. In Sect. 4 we will present a model for the dependence of the strength on time and temperature. In this model a constant shear modulus g is used, thus $\tau_0 = g\gamma_b$.

2.5.2
Estimates of the Shear Modulus, Shear Strength and Ultimate Strength of Cellulose II, PpPTA and PIPD-HT Fibres

In this section some estimates are given of g, τ_0 and σ_L using the lattice constants of the crystal structures, the debonding energy $W_m{}^S$ and results from the molecular force field calculations. In principle, the estimated value for σ_L refers to a fibre with perfect orientation of the chains parallel to the fibre axis and free from impurities. As a lower bound for W_m^S the observed activation energy per mole, W_a, obtained from creep measurements is used. It is also possible to calculate with Eq. 47 the ultimate strength σ_L based on the observed fracture energy W_b. This so-called "observed" ultimate strength, σ_L^{obs}, is then the maximum strength of a domain in the real fibre corresponding to the observed fracture envelope.

2.5.2.1
Cellulose II

In cellulose II with a chain modulus of 88 GPa the likely shear planes are the 110 and 020 lattice planes, both with a spacing of d_c=0.41 nm [26]. The periodic spacing of the force centres in the shear direction along the chain axis is the distance between the interchain hydrogen bonds $p=c/2$=0.51 nm (c chain axis). There are four monomers in the unit cell with a volume V_{cell}=68·10^{-30} m^3. The activation energy for creep of rayon yarns has been determined by Halsey et al. [37]. They found at a relative humidity (RH) of 57% that W_a=86.6 kJ mole^{-1}, at an RH of 4% W_a=97.5 kJ mole^{-1} and at an RH of <0.5% W_a=102.5 kJ mole^{-1}. Extrapolation to an RH of 65% gives W_a=86 kJ mole^{-1} (the molar volume of cellulose taken by Halsey in his model for creep is equal to the volume of the unit cell instead of one fourth thereof).

Two cases will be discussed, viz. the bone-dry cellulose II and the fibre conditioned at 65% RH. For bone-dry cellulose the activation energy per unit volume, $W^S=W_a/(N_A V_{cell})$, is 0.25 GJ m^{-3}. Equations 55 and 54 give g=6.4 and τ_0=1.27 GPa, respectively, with τ_0 being 20% of g. Measurement of the dynamic modulus of a thoroughly dried highly oriented fibre (Fibre 1 in Fig. 8) at 20 °C yielded 66 GPa, which results in an internal shear modulus of about 5 GPa [2, 5]. Equation 57 yields the ultimate strength σ_L=5.3 GPa. For conditioned cellulose the activation energy per unit volume is W^S=0.21 GJ m^{-3}. Equation 55 results in g=5.4 GPa, which lies somewhat above the range of 2.1–3.8 GPa determined from measurements of the modulus at sonic frequencies of various cellulose II fibres conditioned at an RH of 65%. Equation 57 yields σ_L=4.9 GPa.

The ultimate strength, $\sigma_L{}^{obs}$, of a conditioned fibre can be estimated using the observed strain energy. The strain energy up to fracture of conditioned cellulose II fibres is 0.058 GJ m^{-3}. From Eqs. 55 and 54 the values g=1.5 and τ_b=0.42 GPa are derived, respectively. Equation 57 yields σ_L=2.6 GPa, which should be compared with 1.7 GPa being the strength measured for Fibre 1 in

Fig. 8 at an RH of 65%. This difference demonstrates the channelling effect of the orientation distribution during a tensile test.

2.5.2.2
PpPTA

The weakest bonding in the crystal structure of PpPTA is the van der Waals bonding between the hydrogen-bonded planes [38]. Hence, the most likely shear plane is the 200 plane. However, one can argue that for the failure of PpPTA fibres the hydrogen bonds between the chains also have to be broken. Employing molecular mechanics and accurate force fields for hydrogen bonds, Kooijman and Batenburg calculated a shear modulus of 4.07±0.07 GPa and a fracture shear force of 0.23 nN for the hydrogen bond in a perfectly ordered arrangement of PpPTA chains [38, 39]. The cross-sectional area of the hydrogen bond in PpPTA having an orthogonal unit cell with dimensions a=0.78, b=0.52 and c (chain axis)=1.29 nm is equal to $1/2ac$=0.50 (nm)^2, yielding a shear strength of τ_0=0.46 GPa. Applying the maximum shear strain of the Frenkel model, $\gamma_b=p/(4d_c)$=0.31, whereas according to Yoon's approximation $\gamma_b=\tau_0/g$=0.11 is found. These values render some support for the estimate $\tau_0\approx0.2g$. Equation 45 results in σ_L=4.7 GPa, which is larger than the highest observed strength of 4.5 GPa. The observed strain energy of a PpPTA fibre is about 0.1 GJ m^{-3}, which results with Eq. 47 in σ_L^{obs}=7.8 GPa.

The observed range of the shear modulus varies between 1.5 GPa in filaments of regular count to 3 GPa in microfilaments, which correlates with the degree of orientation and crystalline perfection in the fibres [40]. Compared to the theoretical value of the modulus of shear between two hydrogen-bonded chains of 4.1 GPa, it indicates softening due to the van der Waals bonding between the hydrogen-bonded planes.

2.5.2.3
PIPD (or M5)

There are two shear plane systems in PIPD-HT, viz. the 110 and the $1\bar{1}0$ planes, both having a lattice spacing of 0.32 nm [41–44]. Assuming that the interchain hydrogen bonds have to be broken at fibre fracture, the periodic distance of these bonds along the shear direction is $p=c/2$, where c is the monomer length. According to Hageman et al. a theoretical estimate of the total interchain bonding energy per monomer is 1.57 eV or 152 kJ mole^{-1} [45]. As the volume of a single monomer is $250\cdot10^{-30}$ m^3, the interchain energy per unit volume is 1.0 GJ m^{-3}. So with d_c=0.32 nm, p=0.6 nm and W_m^S=1.0 GJ m^{-3} Eq. 55 yields g=11.4 GPa and with Eq. 54 τ_0=3.4 GPa, which is nearly 30% of g. With e_c=550 GPa Eq. 57 yields σ_L=27 GPa. As for these estimates, a theoretical value for the interchain energy was used; the ultimate strength refers to debonding of every single chain at the moment of fracture. Because lattice distortions and inhomogeneities have little effect on the elastic constants, we conclude from the observed value for g, which varies between 5 and 7 GPa, that the theoretical value of the interchain energy is probably too large by a factor of 2. The present modulus of the PIPD-HT fibre is 300 GPa and the highest observed strength 6.6 GPa for a test length of 10 cm, thus W_b=0.094 GJ m^{-3}. From Eq. 47 the ultimate strength corresponding with the observed tensile curve becomes σ_L^{obs}=11.5 GPa.

2.5.3
Theoretical and Experimental Relations between Strength and Modulus of PpPTA, PBO, PIPD-HT, POK, Cellulose II and PET Fibres

By using the relation for the critical shear strength $\tau_b=2\beta g$, Eqs. 10 and 46, and the assumption that $g \ll e_c$, it can be shown that for well-oriented fibres Eqs. 42 and 43 yield the following relation between the initial modulus and the strength

$$\frac{1}{E} = \frac{1}{e_c} + \frac{\beta^2 (2g + \sigma_b)^2}{2g\sigma_b^2} - \frac{(2g + \sigma_b)^2}{10.4 g^2 e_c} \quad (58)$$

Figure 22 shows this theoretical relation between the strength and the modulus of PpPTA fibres for three values of β. A rapid increase in strength of PpPTA is observed in the modulus range 40<E<100 GPa; for higher moduli the strength levels off. This approach does not take into account that during first loading of the fibre, viscoelastic and plastic deformation contribute to the shear deformation [1, 2, 6]. Therefore, a slightly different method has been applied for the

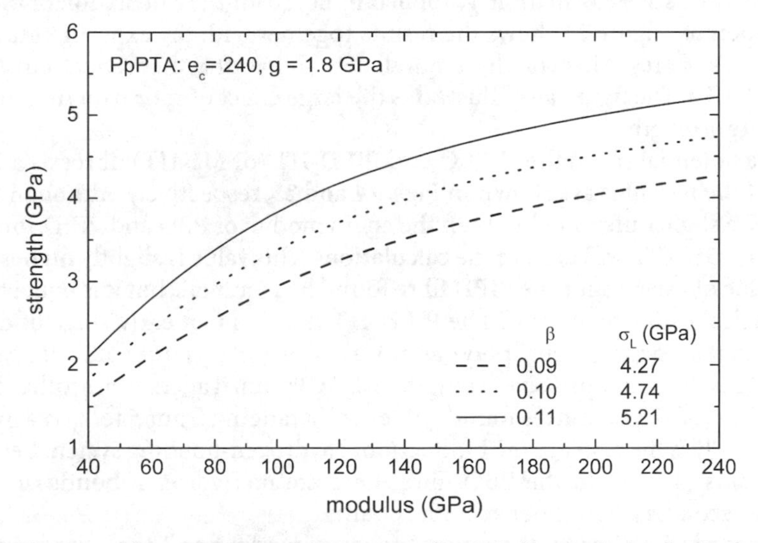

Fig. 22 The strength versus the modulus curves of PpPTA fibres calculated with Eq. 58 for three different values of the critical shear strain

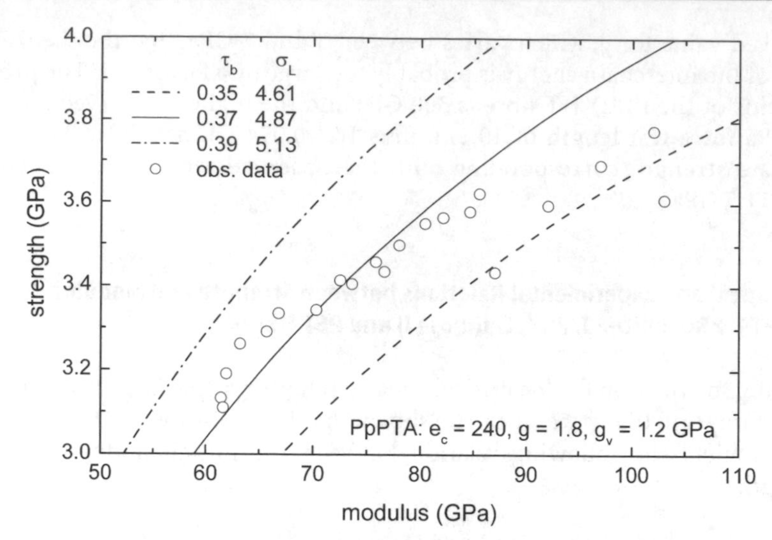

Fig. 23 PpPTA yarn data showing the relation between strength and modulus together with the calculated curves

calculation of the strength as a function of the modulus for a fibre with a single orientation angle. With the estimates for σ_L and τ_b, the strength is calculated for a series of fracture angles in the range $0 \leq \theta_b < \pi/4$ using Eqs. 42 and 43. Next, the initial angle at zero load is obtained from Eq. 26, which permits calculation of the initial modulus with Eq. 10. When the fibres show some viscoelastic and plastic behaviour the shear modulus used in Eq. 26 is smaller (g_v) than the value for purely elastic deformation, g, applied in Eq. 10 for the calculation of the initial modulus. Figure 23 shows the results together with the experimental data on PpPTA fibres. Most of the data closely follow the theoretical curve for τ_b=0.37 GPa. The figure also illustrates the large effect of τ_b or β on the value of the fibre strength.

The potential strengths of PBO and PIPD-HT (or M5-HT) fibres as a function of the modulus are shown in Figs. 24 and 25, respectively. Although there may be a slight difference between the chain moduli of PBO and PIPD, for both fibres e_c=550 GPa was used in the calculations. This value is slightly higher than the value of 496 GPa for the PIPD fibre found by Lammers, but it is equal to the theoretical estimate [42–44]. The PIPD-HT fibre is in an early stage of development. It is an exceptional polymer fibre because of its large value of g, which results in a high compressive strength of 1.7 GPa, unrivalled by any other polymer fibre [8, 42]. Lammers found values for g ranging from 5 to 7 GPa, which presumably is the result of the bidirectional hydrogen-bonding system between the chains [42, 43]. In the PBO fibre there are no hydrogen bonds or other strong secondary bonds between the chains.

Theoretical estimates show that the π–π bonding and the other non-hydrogen-bonding energy contributions in the PIPD-HT fibre constitute about

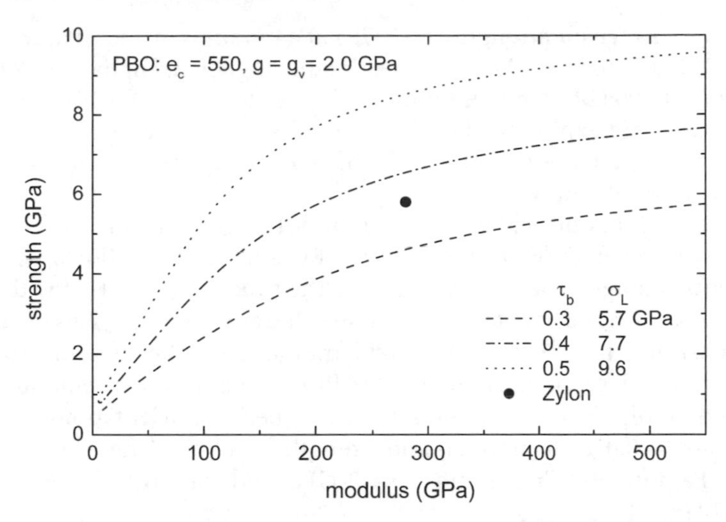

Fig. 24 The strength versus the modulus curves for PBO fibres calculated for three different critical shear stress values and the observed strength of PBO (Zylon) given by the manufacturer

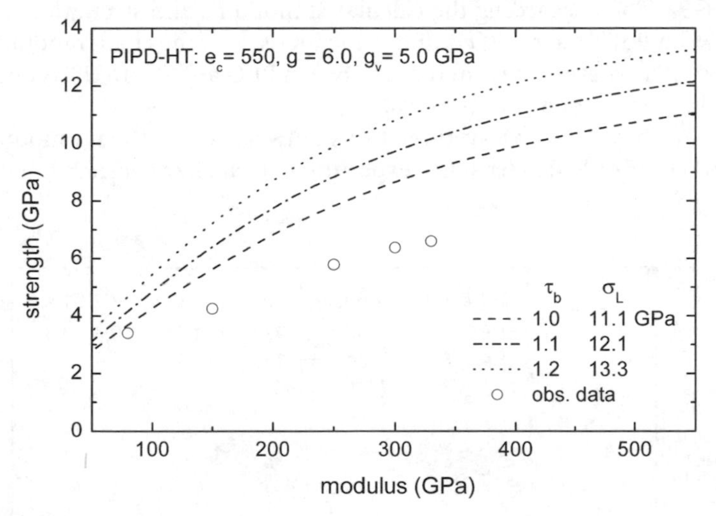

Fig. 25 The strength versus the modulus curves for PIPD-HT (or M5-HT) fibres calculated for three different critical shear stress values and the observed values

35% to the total interchain energy including the hydrogen bonds. This gives some indication for the value of the internal shear modulus g in PBO. Attempts to determine g yielded values between 1 and 4 GPa (unpublished results). Due to the considerably lower value of g in PBO fibres as compared with PIPD-HT, the increase in strength with the modulus occurs in PBO more rapidly than in PIPD-HT. By using for τ_b a value equal to 20% of g we find in the case of PBO for τ_b=0.4 GPa a tensile strength of 6.8 GPa at a modulus of 300 GPa, and in the case of PIPD-HT for τ_b=1.2 GPa a considerably higher strength of 10.7 GPa at a modulus of 300 GPa. On these grounds higher tenacities for PIPD-HT than for PBO fibres are expected. The maximum observed filament properties at a test length of 10 cm are E=330 and σ_L=6.6 GPa for PIPD-HT, and E=260 and σ_L=7.3 GPa for PBO fibres.

Figures 26 and 27 present the modulus and strength as a function of the orientation parameter $\sin^2\Theta$ for the PpPTA, PBO and PIPD-HT fibres, assuming a fibre with a single orientation angle. As the precise value of g for the PBO fibre is not known, we have taken the same value as for PpPTA. This enables us to demonstrate the effect of the chain modulus on the modulus and the strength of the fibres, in particular at medium values of the orientation parameter and for highly oriented fibres. For example, at an orientation parameter value of $\sin^2\Theta$=0.028 the modulus and strength for a PpPTA fibre are E=84 and σ_b=3.9 GPa, for PBO E=104 and σ_b=5.2 GPa, and for PIPD-HT E=241 and σ_b=8.6 GPa.

When the orientation in the fibre is improved, yielding an initial orientation parameter of $\sin^2\Theta$=0.006, the theoretical results are for the PpPTA fibre E=171 and σ_b=4.95 GPa, for PBO E=287 and σ_b=7.3 GPa, and for PIPD-HT E=431 and σ_b=11.2 GPa. Thus, regarding the calculated modulus and strength the difference between PpPTA and PBO is due to the increase of the chain modulus from 240 to 550 GPa, whereas the difference between PBO and PIPD-HT is due to the increase in g from about 1.8 to 6 GPa.

Figure 28 shows the observed relations between strength and modulus for polyetherketone or POK yarns. The experimental data have been obtained from

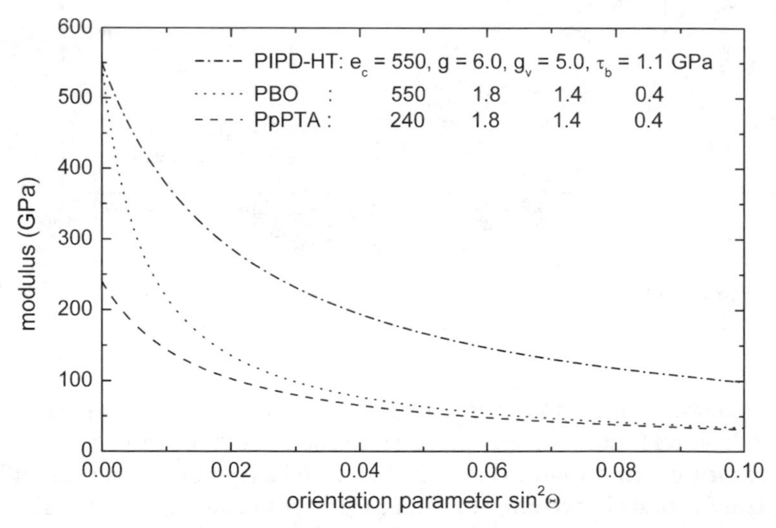

Fig. 26 The modulus as a function of the initial orientation parameter for PIPD-HT, PBO and PpPTA fibres

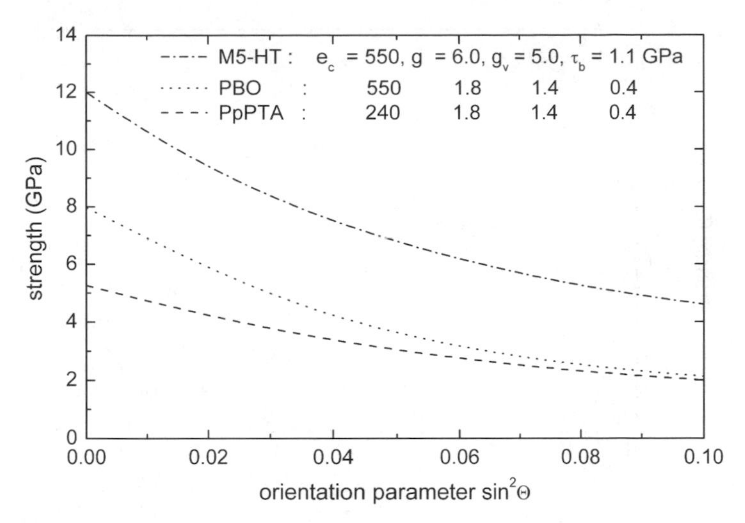

Fig. 27 The strength as a function of the initial orientation parameter for PIPD-HT, PBO and PpPTA fibres

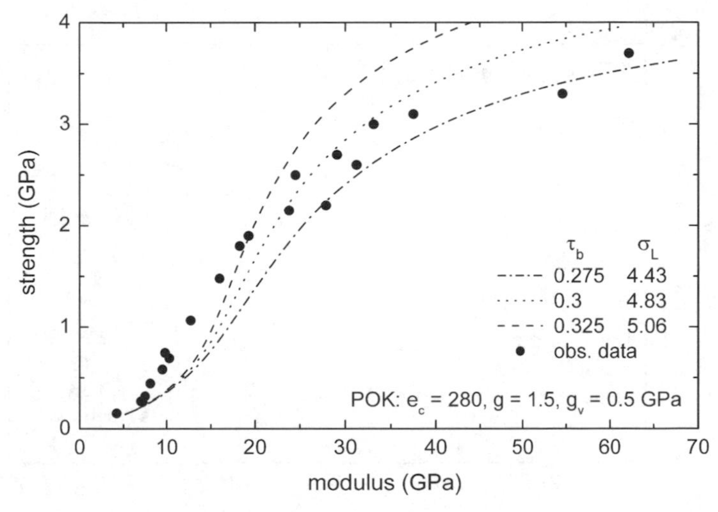

Fig. 28 The observed strength as a function of the initial modulus for a series of POK yarns made by melt spinning and by gel spinning compared with the calculated curves

yarns spun by two different processes, viz. melt spinning and gel spinning [27–29]. Figure 29 shows these relations for cellulose II fibres. The experimental data are for single filaments taken from a single yarn bundle of cellulose II spun from a liquid crystalline solution [26]. The data show a considerable spread of the modulus of the individual filaments due to slight differences of the effective draw ratio in the spinning process. Because the spread of the

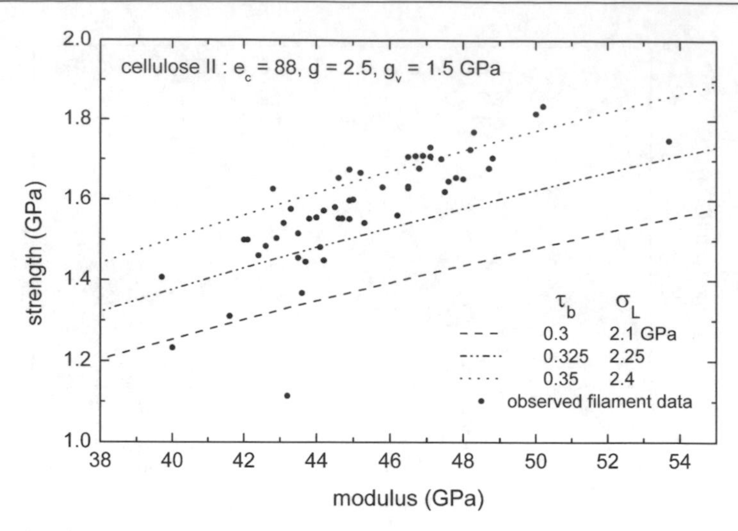

Fig. 29 The observed strength as a function of the initial modulus of filaments taken from a single yarn of cellulose II spun from a liquid crystalline solution compared with the calculated curves [26]

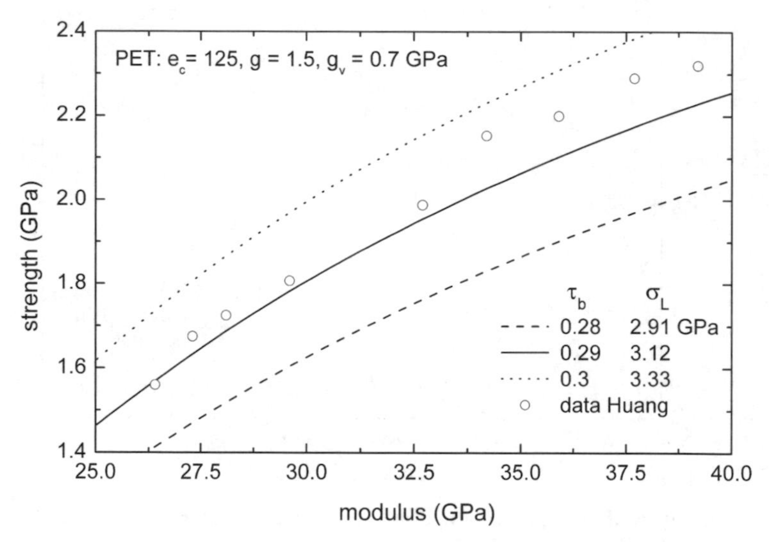

Fig. 30 The observed strength as a function of the modulus for high molecular weight PET yarns made by Huang et al. compared with the calculated curves [46]

strength of these filaments is almost completely explained by the theoretical relation between modulus and strength, it can be concluded that the spinning process is nearly optimised. Figure 30 shows the calculated curves and the observed data for PET fibres [46]. For these three different fibres the agreement between the theoretical curves and the observed data is surprisingly good.

2.6 Relation between the Concentration of the Spinning Solution and the Strength

For the wet-spinning process of lyotropic solutions of polymers using air-gap technology, Picken has derived a relationship between the modulus and the concentration of the spinning solution [47–51]. As shown earlier in this chapter, an increase of the modulus results in an increase of the strength of the fibre. Hence, there should be a relationship between the strength and the concentration. First the relationship between the chain orientation and the concentration of the spinning solution is presented. The unoriented spinning solution is pictured to consist of a randomly oriented collection of anisotropic domains, each with an internal order parameter

$$\langle P_2 \rangle = 1 - \frac{3}{2} \langle \sin^2 \Theta \rangle \tag{59}$$

This equation describes the orientation around a common axis called the director of the domain. For perfectly parallel orientation $\langle P_2 \rangle$ equals 1. The orientation of the directors in the solution is described by the order parameter P_D. The overall orientational order of the anisotropic solution is given by

$$\langle \overline{P_2} \rangle = P_D \cdot \langle P_2 \rangle \tag{60}$$

During spinning the domains are oriented by shear flow just before and in the spinneret, and by the elongational flow in the air gap. For high draw ratios it can be shown that $P_D \rightarrow 1$, with the result that the degree of orientation in the fibre is approximately equal to $\langle P_2 \rangle$, i.e. the internal orientational order in the domain. The internal order parameter $\langle P_2 \rangle$ is determined by the concentration of the solution, the molecular weight, the persistence length of the chain and the temperature of the coagulation bath. The relation between the internal order parameter and the concentration of the spinning solution is derived from the extended Maier–Saupe mean field theory and can be conveniently calculated with an accuracy of about 1% using the equations

$$\langle P_2 \rangle = 0.1 + 0.9 \left[1 - 0.99 \left(\frac{T}{T_{ni}} \right)^3 \right]^{1/4} \quad \text{for} \quad T > T_{ni} \tag{61a}$$

$$\langle P_2 \rangle = 0 \quad \text{for} \quad T \leq T_{ni} \tag{61b}$$

with

$$T_{ni} \approx B c^{\alpha} \tag{62}$$

where T_{ni} is the nematic–isotropic transition temperature in Kelvin, c the concentration in wt.% and T the temperature of the coagulation bath. For PpPTA the parameters are B=76 K and α=0.66. The value 3 of the exponent in Eq. 61a for the ratio T/T_{ni} results from the extension of the theory to polymers and

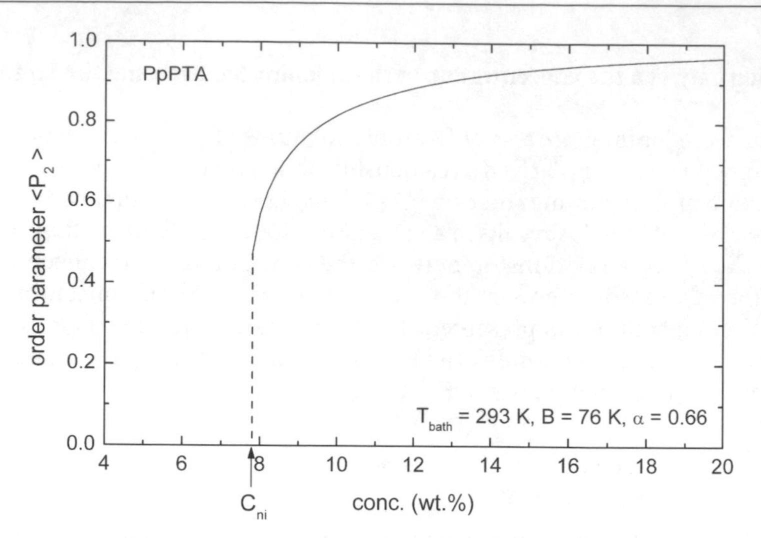

Fig. 31 The order parameter $\langle P_2\rangle$ as a function of concentration for DABT and PpPTA

holds for the whole concentration and temperature range [48–51]. Figure 31 shows $\langle P_2\rangle$ as a function of the concentration. Note that even for the maximum possible concentration (c=20%) of the PpPTA solution at 20 °C the order parameter is 0.964 or $\langle \sin^2\Theta\rangle$=0.024. This implies that the maximum value of the fibre modulus that can be obtained by spinning (i.e. without post-drawing of the fibre) is about 110 GPa, using g=2.5 and e_c=240 GPa. Lowering the bath temperature, increasing the molecular weight or applying a smaller spinneret that diminishes the effect of orientation relaxation during coagulation will increase this value. For example, for a bath temperature T_{bath}=1 °C, while keeping the other conditions unchanged, the order parameter calculated with Eq. 61a becomes 0.971, implying a maximum modulus of 124 GPa.

Similar to the calculation of the fibre strength as a function of the initial modulus, we can now compute the fibre strength as a function of the concentration of the spinning solution [51]. First the fibre strength is calculated for a series of fracture angles θ_b using Eqs. 42, 43 and 45; next, the corresponding series of initial orientation angles is calculated with Eq. 26. Subsequently, the concentration is computed with Eqs. 59, 61a and 62. Note that in this simplified approach a single chain angle is used instead of an orientation distribution. The result for PpPTA fibres is depicted in Fig. 32. This figure shows that in the concentration range $c_{\mathrm{ni}}<c<20\%$ the strength of PpPTA fibres strongly depends on the degree of chain orientation in the spinning solution, being a function of the molecular weight, the persistence length of the polymer chain, the temperature and the concentration of the spinning solution.

The proposed relation between the strength and the modulus of the fibre and the relation between the degree of chain orientation and the concentration of the spinning solution yield a relation between strength and concentration

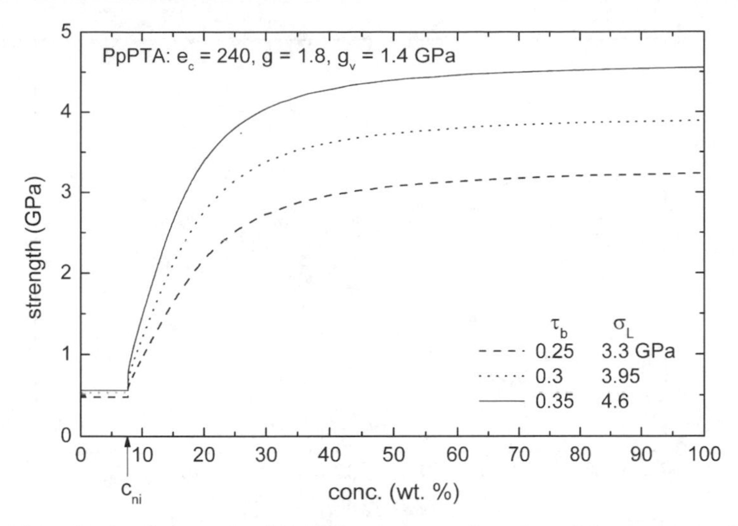

Fig. 32 The calculated strength of PpPTA yarns as a function of the concentration. The isotropic–nematic concentration is indicated by c_{ni}

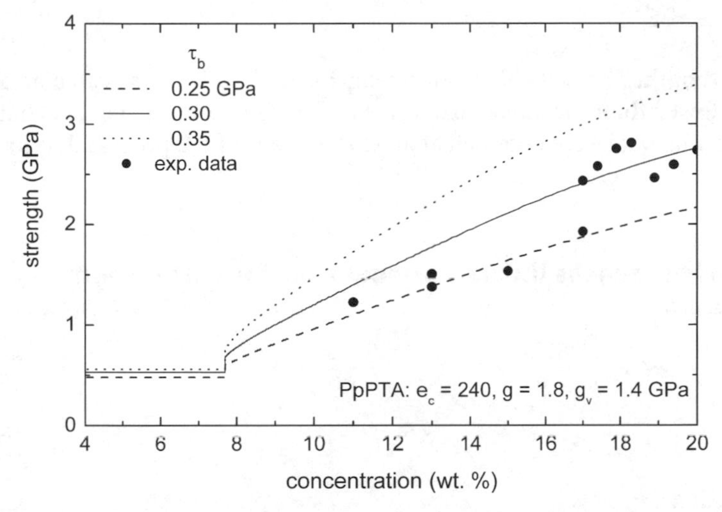

Fig. 33 The observed data for PpPTA yarns showing the relation between the strength and the concentration of the spinning solutions. *Drawn lines* have been calculated with Eqs. 26, 42, 43, 45, 59, 61a, 61b and 62

that can easily be tested. Figures 33 and 34 compare the theoretical curves between the strength and the concentration with the experimental data of PpPTA and poly(*p*-benzanilide terephthalamide), or DABT, respectively. The data confirm that high-strength fibres are obtained by spinning at high concentrations of the spinning solution. Figure 33 suggests that at high concentrations the orientational order of the PpPTA fibre improves, resulting in a higher critical

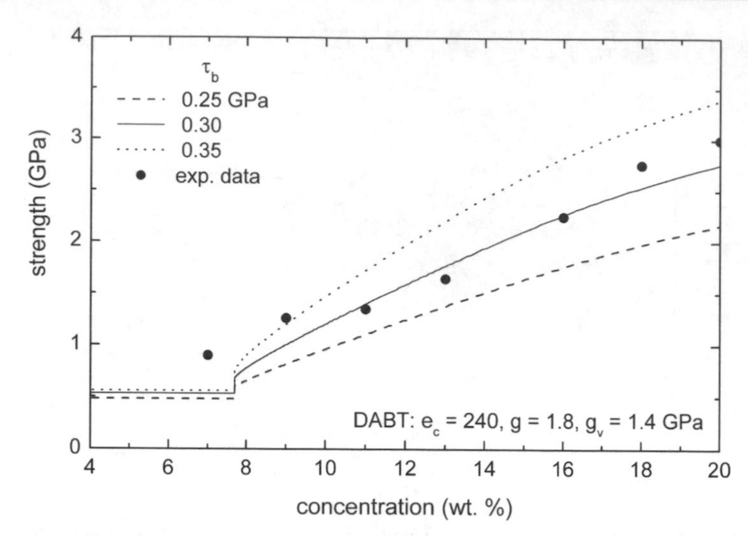

Fig. 34 The observed data for DABT yarns showing the relation between the strength and the concentration of the spinning solutions. *Drawn lines* have been calculated with Eqs. 26, 42, 43, 45, 59, 61a, 61b and 62

shear strength. The critical shear strengths applied in these calculations are slightly lower than the values used in Fig. 23. They reflect the fact that these experimental data were obtained at an earlier stage of the process development.

3 Relation between the Ultimate Strength and the Chain Length Distribution

3.1 Theory

3.1.1 Introduction

The upper limit of the strength in the molecular composite model presented in Sect. 2 is given by Yoon's equation (Eq. 45) [11]. The strength model for polymer fibres developed by Yoon is based on Piggott's model of the strength of uniaxially oriented short-fibre-reinforced macrocomposites [35]. This shear lag model provides good predictions of the strength of macrocomposites, including systems containing low as well as high volume fractions of fibres. We briefly cite here the general outline of Yoon's model: "The fibres are considered an assemblage of fully extended, rod-like polymer chains. The polymer chains are approximated as linear rods of infinite strength under tensile loads. The length of the polymer chains follows the most probable distribution since these polymers were prepared by condensation polymerisation. A macroscopic load is applied to both ends of the fibre and the load is distributed on each polymer chain essentially by the action of interchain forces. The interchain forces are relatively weak and the interchain bonding breaks whenever the local interchain load exceeds a critical value. Although it may not be rigorously justifiable to apply the macroscopic kinematics down to the molecular scale, the situation of the fibre closely resembles the short-fibre composite with a debonding interface. Therefore, the theory is similar in development to the shear lag model of fibre-reinforced composites."

In the macrocomposite model it is assumed that the load transfer between the rod and the matrix is brought about by shear stresses in the matrix–fibre interface [35]. When the interfacial shear stress exceeds a critical value τ_0, the rod debonds from the matrix and the composite fails under tension. The important parameters in this model are the aspect ratio of the rod, the ratio between the shear modulus of the matrix and the tensile modulus of the rod, the volume fraction of rods, and the critical shear stress. As the chains are assumed to have an infinite tensile strength, the tensile fracture of the fibres is not caused by the breaking of the chains, but only by exceeding a critical shear stress. Furthermore, it should be realised that the theory is approximate, because the stress transfer across the chain ends and the stress concentrations are neglected. These effects will be unimportant for an aspect ratio of the rod $L/d>10$ [35].

For the transformation of the macrocomposite model to a molecular composite model for the ultimate strength of the fibre the following assumptions are made: (1) the rods in the macrocomposite are replaced by the parallel-oriented polymer chains or by larger entities like bundles of chains forming fibrils; and (2) the function of the matrix in the composite, in particular the rod–matrix interface, is taken over by the intermolecular bonds between the chains or fibrils. In order to evaluate the effect of the chain length distribution on the ultimate strength the monodisperse distribution, the Flory distribution, the half-Gauss and the uniform distribution are considered.

It is assumed that all chains are oriented parallel to the fibre axis and that equal parts of the fibre contain equal numbers of starting points of chains of arbitrary length u. Thus the starting points are distributed homogeneously along the length of the fibre. The length distribution of the chains is determined by the distribution function $f(u)$ with

$$c_0 \int_{u_0}^{\infty} f(u)\,du = 1 \tag{63}$$

where u_0 is the length of the monomer in metres and c_0 a normalisation constant with dimension m^{-1}. It is assumed that there are N homogeneously distributed starting points per unit length of the fibre. So in a fibre part of length u there are Nu starting points of chains. The number of chains, dn, with a length between u and $u+du$, follows from the length distribution function $f(u)$

$$dn = c_0 N u f(u)\,du \tag{64}$$

Each of these dn chains of length u passes through the cross section of the fibre and there are no other chains of length u that pass through this cross section. The total number of chains passing through a cross section is given by

$$M = c_0 N \int_{u_0}^{\infty} u f(u)\, du = c_0 N u_a \tag{65}$$

where u_a is the average chain length.

At this early stage we present a modification of Yoon's model by introducing the crossing length distribution. This probability function $h(u)$ of the lengths of chains which pass through a cross section of the filament is defined by

$$h(u) = \frac{c_0 N u f(u)}{M} = \frac{u}{u_a} f(u) \tag{66}$$

By means of Eq. 66 it is expressed that long chains have a larger probability to cross a fibre cross section than short chains. Yoon does not make a distinction between $f(u)$ and $h(u)$.

The number of chains M that pass through a cross section of the fibre can be expressed in the chain volume V_c of the fibre. Let the cross section of a chain be A_c, then the N starting points of chains per unit length of the fibre yield a volume contribution $c_0 N u_a A_c$. The chain volume in a part of the fibre with length L and cross section D becomes $c_0 N u_a A_c L$, which gives a chain volume fraction

$$V_c = \frac{c_0 N u_a A_c}{D} \tag{67}$$

Equations 65 and 67 yield

$$M = c_0 N u_a = \frac{V_c D}{A_c} \tag{68}$$

3.1.2
Forces Acting on the Chain

For a chain of length u which is bonded by means of intermolecular forces over the complete length, the Piggott force model for the tensile stress σ in the chain as a function of the position y from the centre of the chain is given by

$$\sigma(y) = e_c \varepsilon \left[1 - \frac{\cosh(2ay/u_0)}{\cosh(au/u_0)} \right] \quad \text{for} \quad 0 \le |y| \le \tfrac{1}{2} u \tag{69}$$

with e_c the chain modulus, ε the strain of the fibre and $a=\mu u_0 (2r)^{-1}$ with r being the chain radius, $\mu^2=4g(Ce_c)^{-1}$, g the shear modulus of the matrix and $C=\log\{2\pi(V_c\sqrt{3})^{-1}\}$ [11]. The shear stress τ at the chain surface is given by

$$\tau(y) = \tfrac{1}{2}\mu e_c \varepsilon \frac{\sinh(2ay/u_0)}{\cosh(au/u_0)} \quad \text{for} \quad 0 \le |y| \le \tfrac{1}{2}u \tag{70}$$

Figures 35 and 36 show σ and τ as a function of the distance y to the chain centre for completely bonded chains of 5, 10 and 15 monomeric units and a fibre strain of ε=0.02 calculated with Eqs. 69 and 70. The tensile stress σ is symmet-

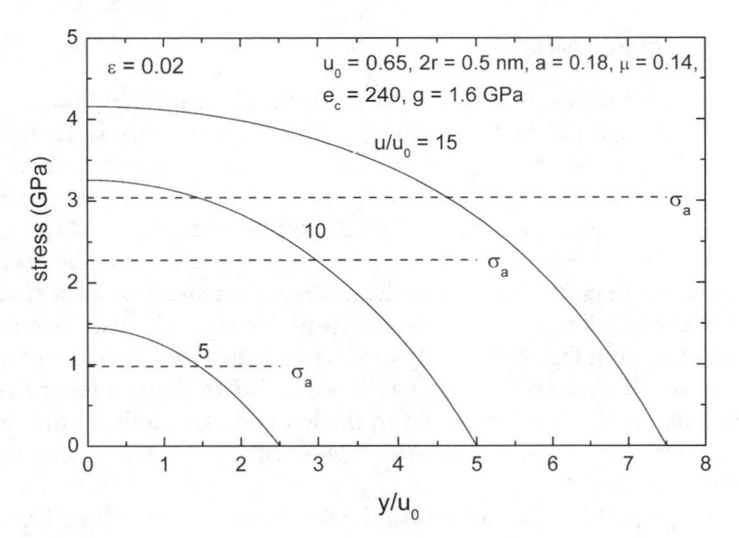

Fig. 35 The tensile stress as a function of the distance y from the chain centre for chains consisting of 5, 10 and 15 units and for a fibre strain ε=0.02

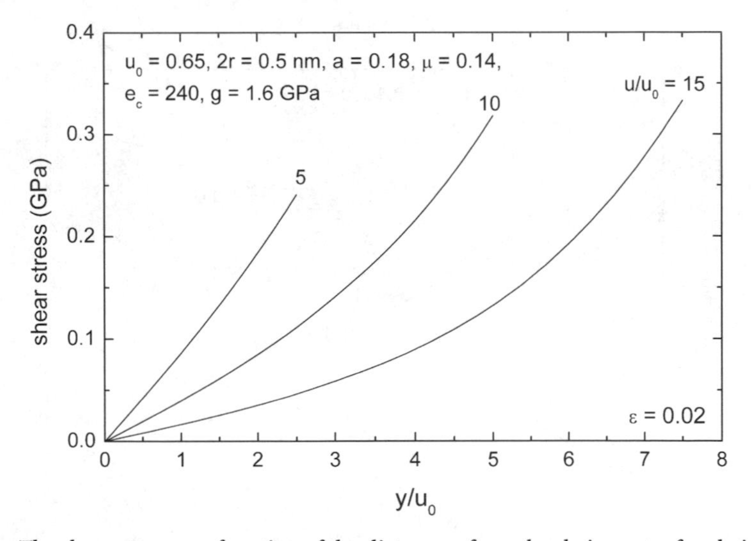

Fig. 36 The shear stress as a function of the distance y from the chain centre for chains consisting of 5, 10 and 15 units and for a fibre strain ε=0.02

rical with respect to the chain centre and the maximum value σ_{max} is reached halfway between the chain ends. The shear stress τ is antisymmetric with respect to the chain centre. The absolute maximum value τ_{max} is reached at the chain ends. The maximum values are given by

$$\sigma_{max} = e_c\varepsilon\left[1 - \frac{1}{\cosh(au/u_0)}\right] \tag{71}$$

$$\tau_{max} = \tfrac{1}{2}\mu e_c\varepsilon \tanh(au/u_0) \tag{72}$$

Apart from the fibre strain ε all quantities in Eqs. 71 and 72 are material dependent. As an example the functions σ_{max} and τ_{max} are drawn in Figs. 37 and 38 for a=0.18, μ=0.14 and ε=0.02 and 0.03 with $\mu e_c\varepsilon$ equal to 0.67 and 1.0, respectively. Further, it is assumed that the critical (debonding) shear stress τ_0 equals 0.4 GPa. For these parameters Figs. 37 and 38 depict the maximum stress and the maximum shear stress as a function of the relative chain length u/u_0. As can be seen in Fig. 38, the critical shear stress on a chain is not reached for ε=0.02, not even for long chain lengths. If the fibre strain is increased to ε=0.03, it can be found from Fig. 38 that only the chains with a length of six monomer units (u/u_0=6) or shorter are completely bonded. The chains longer than six units are only bonded for three units to the left and three units to the right of the chain centre. The maximum tensile stress in such a chain of six units equals 2.7 GPa, see also Fig. 37.

The average tensile stress for a completely bonded chain follows from

$$\sigma_a = \frac{2}{u}\int_0^{u/2} \sigma(y)dy = e_c\varepsilon\left[1 - \frac{\tanh(au/u_0)}{au/u_0}\right] \tag{73}$$

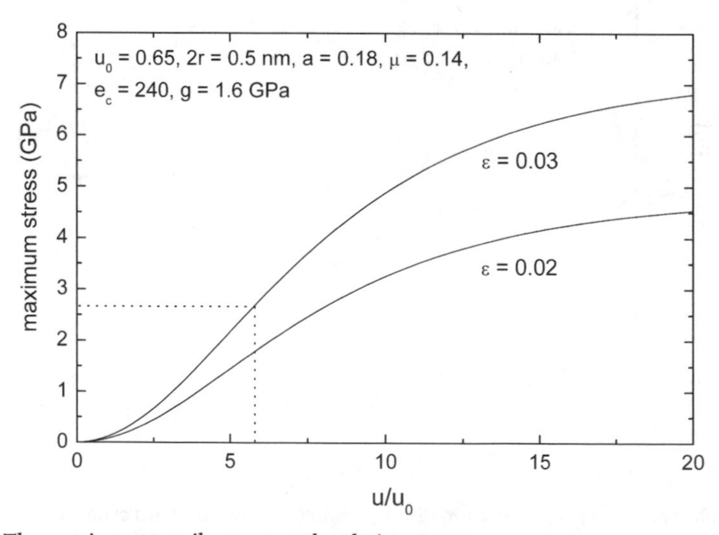

Fig. 37 The maximum tensile stress at the chain centre

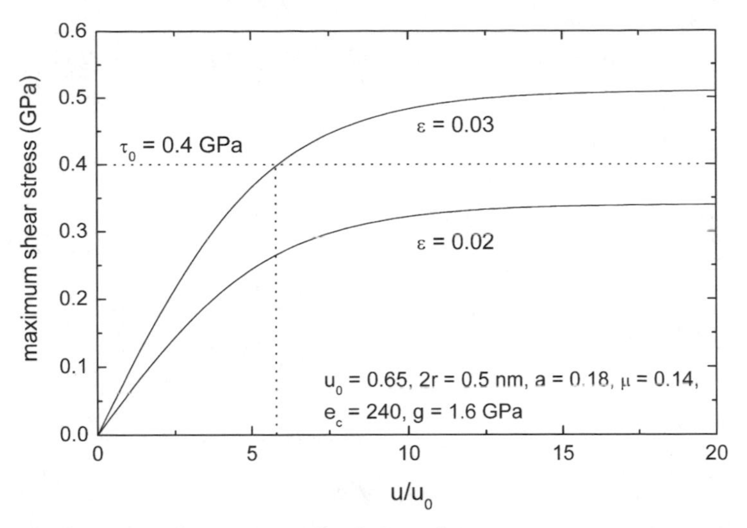

Fig. 38 The maximum shear stress at the chain end

Examples of the average stress are shown in Fig. 35.

For partially bonded chains the tensile and shear stresses become

$$\sigma(y) = e_c\varepsilon\left[1 - \frac{\cosh(2ay/u_0)}{\cosh(au_c/u_0)}\right] \quad \text{for} \quad 0 \le |y| \le \tfrac{1}{2}u_c \tag{74a}$$

$$\sigma(y) = 0 \quad \text{for} \quad \tfrac{1}{2}u_c \le |y| \le \tfrac{1}{2}u \tag{74b}$$

and

$$\tau(y) = \tfrac{1}{2}\mu e_c\varepsilon\frac{\sinh(2ay/u_0)}{\sinh(au_c/u_0)} \quad \text{for} \quad 0 \le |y| \le \tfrac{1}{2}u_c \tag{75a}$$

$$\tau(y) = 0 \quad \text{for} \quad \tfrac{1}{2}u_c \le |y| \le \tfrac{1}{2}u \tag{75b}$$

In Eqs. 74a, 74b, 75a and 75b the quantity u_c equals the bonded length of the chain. These relations are almost equal to Eqs. 69 and 70, except for the bonded length u_c which replaces the total length u. For all chains longer than u_c the stress models are identical to the models for length u_c, and the remaining tails of the chains are stress free because they are debonded. Figures 39 and 40 give examples of the stress distributions for partially debonded chains for a fibre strain of ε=0.03. The average stress in a partially debonded chain is given by

$$\sigma_a = e_c\varepsilon\frac{u_c}{u}\left[1 - \frac{\tanh(au_c/u_0)}{au_c/u_0}\right] \tag{76}$$

As remarked earlier, in the example of Figs. 37 and 38 all chains longer than six units are partially debonded for a fibre strain of ε=0.03. Figure 39 shows that

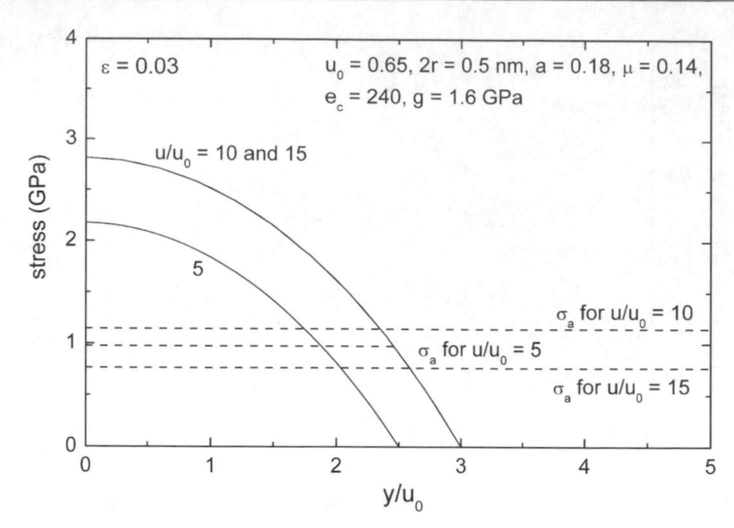

Fig. 39 The tensile stress in a partially debonded chain as a function of the distance y from the chain centre for chains consisting of 5, 10 and 15 units and a fibre strain ε=0.03

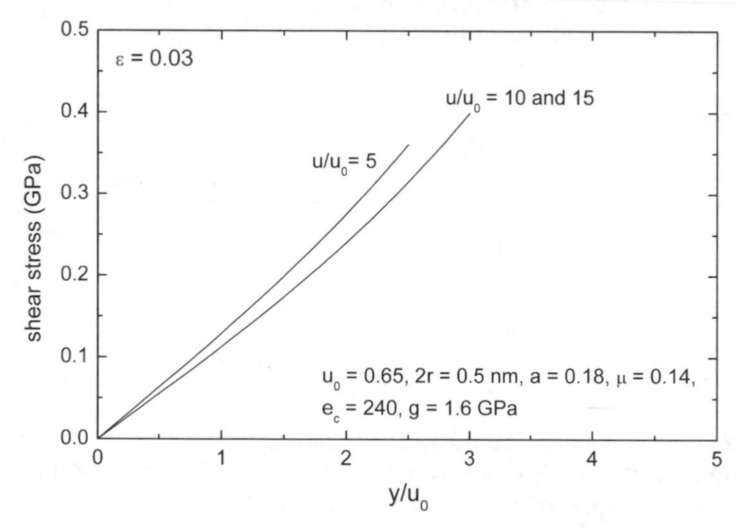

Fig. 40 The shear stress on a partially debonded chain as a function of the distance y from the chain centre for chains consisting of 5, 10 and 15 units and a fibre strain ε=0.03

the stress distributions for a chain of ten units equals the distributions for a chain of 15 units. Both chains behave as a chain of six units. The stress-free or debonded ends on both sides of the chain are two units, respectively 4.5 units long. The average tensile stress for a chain of 15 units is lower than that for a chain of ten units because of the longer stress-free tails. For all chains in the fibre the shear stress has its maximum on the chain end. If a chain end is partially debonded the maximum is reached at the end of the bonded part, for $z=1/2u_c$. For all chains longer than u_c this maximum is the same and equals τ_0 at the end of the bonded part

$$\tau_0 = \tfrac{1}{2}\mu e_c \varepsilon \tanh(au_c/u_0) \tag{77}$$

Except for the fibre strain this critical debonding stress is thus a material constant. The starting points of the chains of length u which pass through a certain fibre cross section are uniformly distributed over the fibre part of length u left from the cross section. This means that the crossing point of chain and fibre cross section is such that the left-hand-side length is uniformly distributed between 0 and u. The expected tensile stress in a chain length u in the crossing point with the fibre cross section then equals σ_a, given by Eqs. 73 and 76 for completely bonded and partially debonded chains, respectively. Besides depending on the material parameters such as e_c, a and u_0, the average tensile stress σ_a also depends on the chain length u, the fibre strain ε and the bonding length u_c. The last two quantities are coupled by means of the material constant τ_0, so the fibre strain can be solved from Eq. 77

$$\varepsilon = \frac{2\tau_0}{\mu e_c} \frac{1}{\tanh(au_c/u_0)} \tag{78}$$

Substitution in Eqs. 73 and 76 gives

$$\sigma_a = \frac{2\tau_0}{\mu \tanh(au_c/u_0)} \left[1 - \frac{\tanh(au/u_0)}{au/u_0}\right] \quad \text{for} \quad u_0 \le u \le u_c \tag{79a}$$

and

$$\sigma_a = \frac{2\tau_0}{\mu \tanh(au_c/u_0)} \frac{u_c}{u} \left[1 - \frac{\tanh(au_c/u_0)}{au_c/u_0}\right] \quad \text{for} \quad u \ge u_c \tag{79b}$$

3.1.3
Force in the Cross Section of the Fibre

The number of chains of length u which cross a certain fibre cross section is determined by the crossing length distribution $h(u)$ defined in Sect. 3.1.1. The tensile stress in a chain passing through a fibre cross section is calculated in Sect. 3.1.2. The force S per unit area in a cross section can be found from

$$S = \frac{MA_c}{D} \int_{u_0}^{\infty} \sigma_a h(u)\, du \tag{80}$$

As long as the fibre strain ε is small enough, all chains are completely bonded and σ_a is given by Eq. 73. Substitution gives

$$S = V_c e_c \varepsilon \int_{u_0}^{\infty} \left[1 - \frac{\tanh(au/u_0)}{au/u_0} \right] h(u)\,du \tag{81}$$

If the chain lenght distribution $f(u)$ of a polymer fibre is known, $h(u)$ can be determined. The monomer length u_0 and the chain volume fraction V_c are also known. The force S is proportional to the chain modulus e_c and the fibre strain. At a certain fibre strain given by Eq. 78, the shear stress on the ends of the longest chains occurring in the chain lenght distribution reach the critical value τ_0. For higher fibre strains the average tensile stress σ_a is given by Eqs. 79a and 79b, and the fibre force S per unit area becomes

$$S = \frac{2V_c\tau_0}{\mu \tanh(au_c/u_0)} \left\{ \int_{u_0}^{u_c} \left[1 - \frac{\tanh(au/u_0)}{au/u_0} \right] h(u)\,du + \int_{u_c}^{\infty} \left[1 - \frac{\tanh(au_c/u_0)}{au_c/u_0} \right] h(u)\,du \right\} \tag{82}$$

Note that the factor within brackets in the second integral is independent of the integration variable u. From Eq. 78 it can be seen that a higher value of the fibre strain ε corresponds with a smaller value of the bonding length u_c. Such a smaller bonding length u_c may give a larger fibre force S per unit area, but this depends on the crossing length distribution $h(u)$, and thus on the molecular weight distribution. It is this particular distribution that determines the contribution of the two different integrals in Eq. 82. If all material quantities and fibre quantities (V_c, τ_0, μ, u_0, a and $h(u)$) are known, S is given as a function of u_c by means of Eq. 82. For a critical value of u_c, which may be infinite, S reaches an absolute maximum S_{max}. This maximum equals the failure stress σ_L, which can be determined in a tensile test of a fibre with chains parallel to the fibre axis; the corresponding value of u_c is the critical bonding length.

The most simple chain lenght distribution is the delta function or the monodisperse distribution $f(u)=\delta(u-u_a)$. All chain lengths are equal to u_a. The maximum of S is reached just prior to debonding, i.e. for $u_c=u_a$

$$S_{max} = \frac{2V_c\tau_0}{\mu \tanh(au_c/u_0)} \left[1 - \frac{\tanh(au_a/u_0)}{au_a/u_0} \right] \tag{83}$$

For a long average chain length u_a, Eq. 83 can be approximated by

$$S_{max} = \frac{2V_c\tau_0}{\mu} \tag{84}$$

The values of S_{max} given by Eqs. 83 and 84 are the same as those given by Yoon, because $h(u)$ equals $f(u)$ for the monodisperse function [11].

3.2
Ultimate Strength for Various Chain Length Distributions

In the examples discussed in this section, the average chain length z_n (or the average degree of polymerisation DP) expressed in the length of the monomeric unit u_0 is based on the number-average length given by

$$z_n = \frac{\int z f(z)\, dz}{\int f(z)\, dz} \tag{85}$$

where z is the chain length in monomeric units (m.u.) and $f(z)$ is the chain length distribution.

The effect of the width of the distribution on the ultimate fibre strength, σ_L, is calculated for the case in which the molecular weight distribution $f_w(z)= zf(z)z_n^{-1}$ is a uniform distribution. The weight-average length of the distribution $f_w(z)$ is defined as

$$z_w = \frac{\int z f_w(z)\, dz}{\int f_w(z)\, dz} \tag{86}$$

Equation 82 is now rewritten in terms of z

$$S = \frac{2V_c\tau_0}{\mu \tanh(az_c)} \left\{ \int_1^{z_c} \left[1 - \frac{\tanh(az)}{az} \right] h(z)\, dz + \int_{z_c}^{\infty} \left[1 - \frac{\tanh(az_c)}{az_c} \right] h(z)\, dz \right\} \tag{87}$$

where z_c is the bonded length of the chain and

$$h(z) = \frac{z}{z_n} f(z) = f_w(z) \tag{88}$$

3.2.1
Monodisperse Distribution

Before we discuss the relation between ultimate fibre strength and degree of polymerisation (z_n or DP), we first show that this model in fact describes the relation between the composite strength and the aspect ratio of the rod. In the case of a monodisperse distribution, i.e. all rods or chains have the same diameter $2r$ and the same length u_a, the ultimate strength as a function of the aspect ratio $b=u_a(2r)^{-1}$ is given by

$$\sigma_L = \frac{2V_c\tau_0}{\mu \tanh(\mu b)} \left[1 - \frac{\tanh(\mu b)}{\mu b} \right] \tag{89}$$

A high mechanical anisotropy is characterised by a small value of μ. Application of the composite model to a fibre implies that V_c=1. For very long chains or $b \rightarrow \infty$ Eq. 45 is derived

$$\sigma_L = \frac{2\tau_0}{\mu} = 1.14\tau_0\sqrt{\frac{e_c}{g}} \tag{90}$$

This expression has been used in Sect. 2. Figure 41 shows the strength as a function of the aspect ratio for two values of the parameter μ. For an aspect ratio of 100 and μ=0.16, the strength is 94% of its maximum value. Comparison of the curves shows that for the higher value of μ the maximum value of the strength is reached earlier than for the lower value of μ, but a stronger mechanical anisotropy results for the same critical shear stress in a higher maximum strength. Thus, a high mechanical anisotropy is advantageous at long chain lengths.

The most obvious building element in polymer fibres is the chain, but other candidates should also be considered. For example, the fibre can be regarded as a composite built up from fibrils. In fact, there are two extremes for the aspect ratio of the building or elemental rod, viz. on the one hand the length of the average chain divided by its radius, and on the other hand the length of the filament itself divided by its radius. In the latter case the diameter of the building rod equals that of the filament diameter, and the fibre or filament may be considered as a polymer whisker. Hence, only in these two cases does one arrive at the largest aspect ratio, viz. the average chain length u_a with $u_a(2r)^{-1}$>100, and

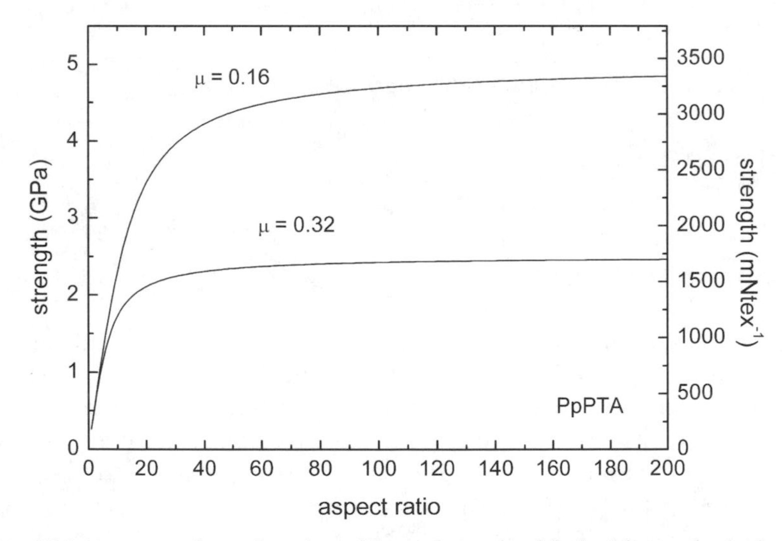

Fig. 41 Ultimate strength as a function of the aspect ratio of the building rod calculated for two values of μ. The shear strength is 0.4 GPa and V_c=1. For PpPTA fibres $\mu \approx 0.16$, using g=2 and e_c=240 GPa

the filament itself with a ratio of test length to filament diameter larger than 100. This suggests that the largest strength might be obtained when any fibrillar structure with cross-sectional diameters between the chain diameter and the filament diameter is absent. So, if fibrils are present, they may take over the role of the rod in the composite model and their aspect ratio may strongly affect the strength of the fibre, the fibril with the largest aspect ratio being the chain itself.

In order to investigate, for a monodisperse distribution of the chain length, the effect of the degree of polymerisation on the fibre strength we write Eq. 89 as

$$\sigma_L = \frac{2V_c\tau_0}{\mu\tanh(az_n)}\left[1 - \frac{\tanh(az_n)}{az_n}\right] \tag{91}$$

where the degree of polymerisation (DP) is given by $z_n=u_a u_0^{-1}$, and $a=\mu u_0(2r)^{-1}$ with u_0 being the length of the monomer. Figure 42 shows the ultimate strength of a PpPTA fibre as a function of the degree of polymerisation for various values of the diameter $2r$, using e_c=240, g=2, τ_0=0.4 GPa and u_0=1.3 nm. As the rod diameter decreases from a value of 10 nm, which is about two times the size of a fibril in a PpPTA fibre, to the value of 0.5 nm pertaining to the cross section of a single chain, the DP range for which the strength is nearly the maximum value of the ultimate strength $\sigma_L=2\tau_0\mu^{-1}$ extends to lower values of the DP. Figure 43 shows a graph of $^{10}\log(\sigma_L)$ versus $^{10}\log(\text{DP})$. Depending on the diameter, the initial part of the curves has a slope of about 0.6 to 0.9. All curves asymptotically reach the value of $^{10}\log(2V_c\tau_0\mu^{-1})$.

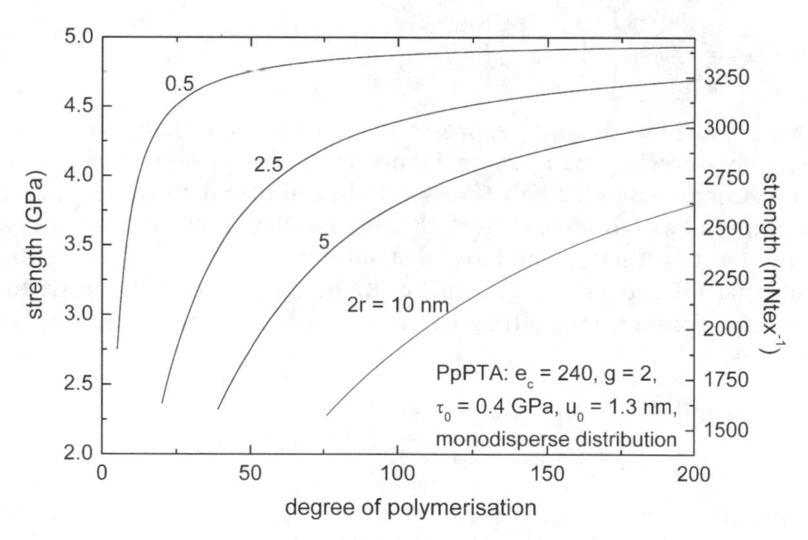

Fig. 42 Ultimate strength σ_L of PpPTA fibre as a function of the degree of polymerisation for a monodisperse distribution calculated for series of diameters $2r$ of the building element, μ=0.16 and for aspect ratios $b=u_a(2r)^{-1}$>10

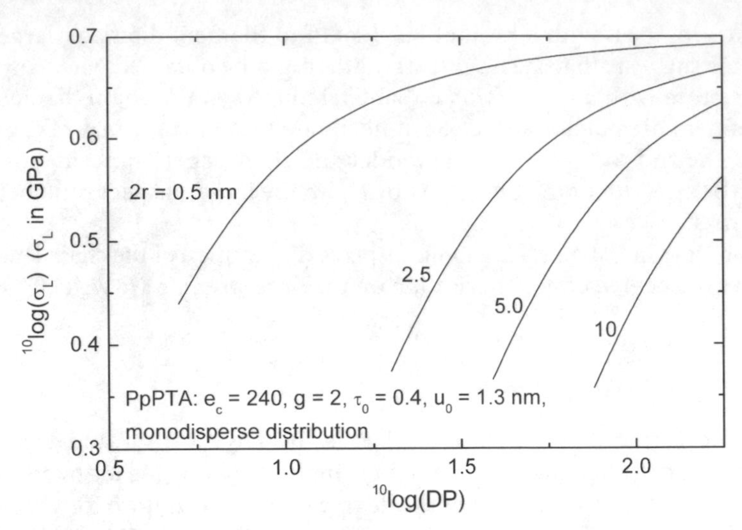

Fig. 43 Double logarithmic graph of the ultimate strength of PpPTA fibre versus the degree of polymerisation for a monodisperse distribution for various values of the diameter and for μ=0.16. Calculation for aspect ratios $b=u_a(2r)^{-1}>10$

3.2.2
Flory Distribution

Polymers prepared by a polycondensation reaction have the Flory distribution for the chain length, which can be approximated by the function

$$f(z) = \frac{1}{z_n - 1} \exp\left[-\left(\frac{z-1}{z_n - 1}\right)\right] \tag{92}$$

where z is the chain length expressed in the monomer length u_0 and z_n is the number-average chain length. Figure 44 shows an example of this distribution. Compared to the half-Gauss and the uniform distribution, the Flory distribution has more very short chains as well as more very long chains. It can be shown that for the Flory distribution $\sigma_L = S_{max}$ for $z_c \rightarrow \infty$, with the result that the second integral in Eq. 87 becomes zero. This leads to the following relation for the ultimate fibre strength as a function of the average DP ($=z_n$)

$$\sigma_L = \frac{2V_c\tau_0}{\mu}\left[1 - \frac{1}{az_n}\int_0^1 \frac{1 - e^{-2a}y^{2a(z_n-1)}}{1 + e^{-2a}y^{2a(z_n-1)}}\,dy\right] \quad \text{for} \quad z_n \rangle (2a)^{-1} + 1 \tag{93}$$

For the same values of e_c, g, τ_0 and u_0 as for the monodisperse distribution, Figs. 45 and 46 show the results for the Flory distribution of chain lengths. The curves start at degrees of polymerisation determined by $z_n=[(2a)^{-1}+1]$. A comparison of Fig. 45 with Fig. 42 shows that, for a diameter equal to the chain di-

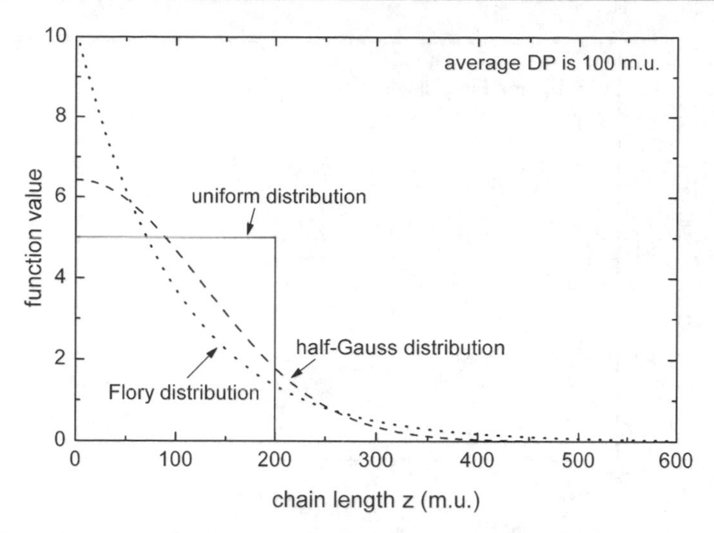

Fig. 44 Flory distribution, half-Gauss and uniform chain length distribution for an average DP of 100 monomeric units (m.u.)

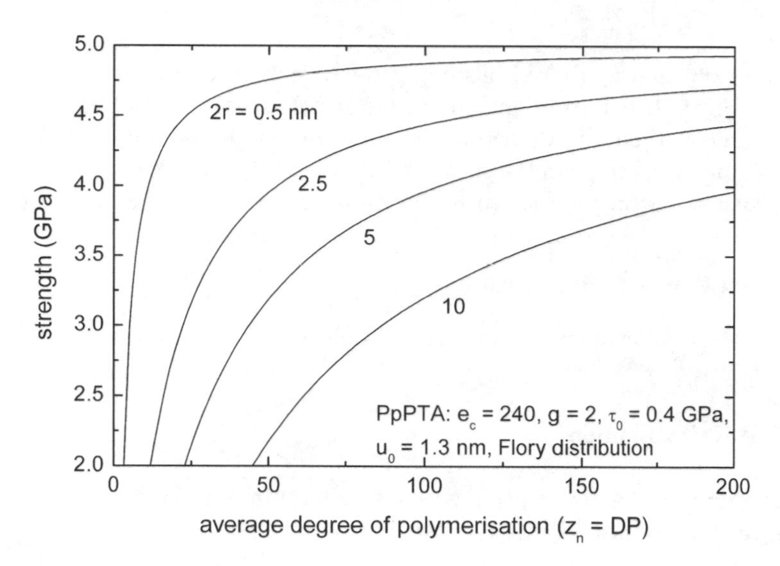

Fig. 45 Ultimate strength of PpPTA fibres versus the degree of polymerisation applying the Flory distribution of chain lengths for various values of the diameter $2r$ calculated with Eq. 93

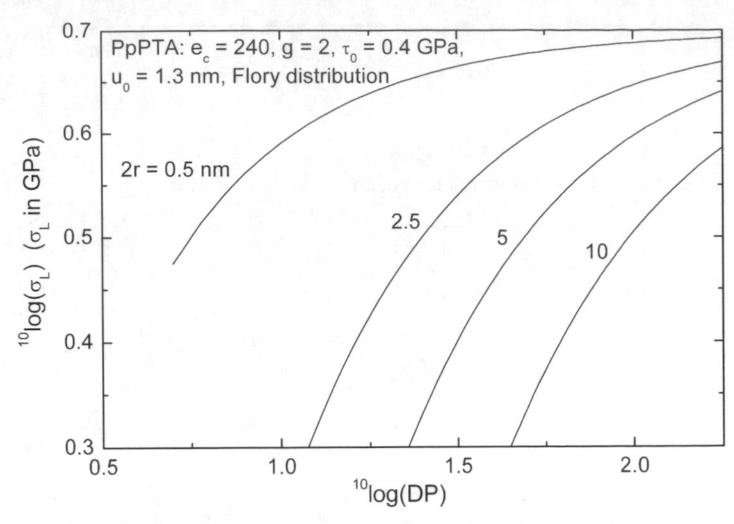

Fig. 46 Double logarithmic graph of the ultimate strength of PpPTA fibre versus the degree of polymerisation applying a Flory distribution of chain lengths calculated with Eq. 93 for various values of the diameter

ameter ($2r$=0.5 nm), there is a small difference between the curves calculated for the Flory distribution and the monodisperse distribution. For increasing rod diameter the Flory distribution results in a higher strength. As shown in Fig. 43, the initial slope in the curves for a monodisperse distribution lies in the range 0.6–0.9, which as shown in Fig. 46 is similar for the Flory distribution, thus

$$\sigma_{\mathrm{L}} \propto (M_{\mathrm{n}})^{\alpha} \qquad \text{with} \qquad 0.6 < \alpha < 0.9 \tag{94}$$

where M_{n} ($=z_{\mathrm{n}}$) is the number-average molecular weight.

3.2.3 Half-Gauss Distribution

As an example of a distribution with a smaller number of long chains than the Flory distribution, a half-Gauss distribution is chosen,

$$f(z) = \frac{2}{\pi(z_{\mathrm{n}} - 1)} \exp\left[-\left(\frac{z-1}{\sqrt{\pi}(z_{\mathrm{n}} - 1)}\right)^2\right] \tag{95}$$

as shown in Fig. 44. The strength versus the degree of polymerisation for PpPTA fibres is presented in Fig. 47. A comparison with the results of Flory distribution presented in Fig. 45 shows that, only for large diameters and lower average chain length, the strength for the half-Gauss distribution is smaller than for the Flory distribution.

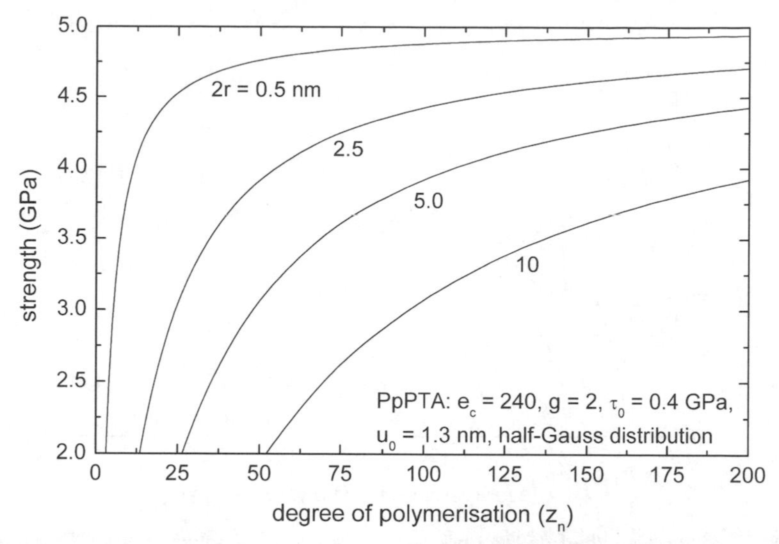

Fig. 47 Ultimate strength of PpPTA fibres versus the degree of polymerisation applying a half-Gauss distribution of chain lengths for various values of the diameter $2r$

3.2.4 Uniform Distribution

To study the effect of the absence of long chains in the distribution the uniform distribution is chosen:

$$f_l(z) = \frac{1}{2(z_n - 1)} \quad \text{for} \quad 1 \le z \le 2z_n - 1 \tag{96a}$$

$$f_l(z) = 0 \quad \text{else} \tag{96b}$$

The results for the PpPTA fibres calculated for this distribution are shown in Fig. 48. For the smallest diameter $2r$=0.5 nm there is, over practically the whole range of the average chain length, no difference with the other chain length distributions. With increasing diameter the strength of the uniform distribution becomes progressively smaller than the strength calculated for the Flory distribution. However, for large values of the average chain length the difference diminishes. Considering the large difference in shape of the chain length distribution selected for the calculation of the ultimate strength curves, the effect on the strength is relatively small.

For investigation of the effect of the width, w, of the distribution on the ultimate fibre strength the uniform chain distribution function is chosen. As in experiments often only the molecular weight distribution $f_w(z)=zf_l(z)z_n^{-1}$ is determined, we used a uniform molecular weight distribution with a width w defined as

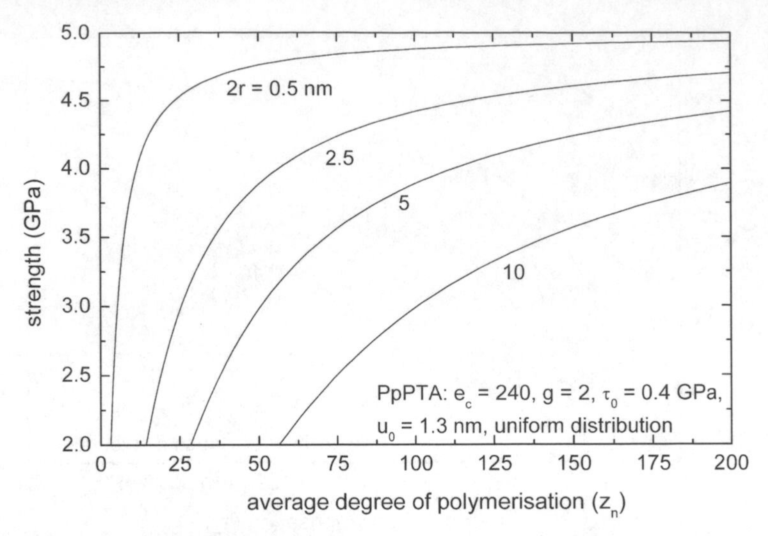

Fig. 48 Ultimate strength of PpPTA fibres versus the degree of polymerisation for a uniform distribution of chain lengths for various values of the diameter $2r$ and with $1 \le z \le 2z_n - 1$

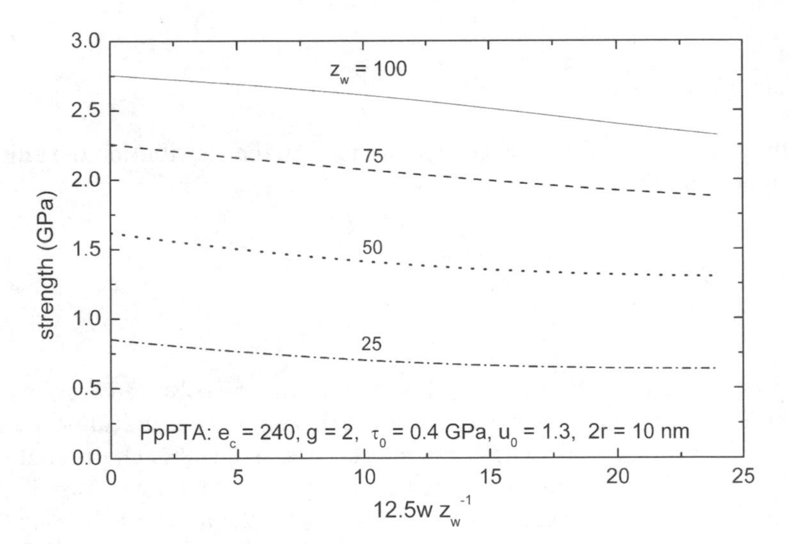

Fig. 49 Effect of the width of a uniform chain length distribution on fibre strength. The width of the block w is centred at the end of the value of the molecular weight, expressed as the length of the monomeric unit. For example, for the point on the curve z_w=100 m.u. and $12.5wz_w^{-1}$=20, it means that the total block width w=160 m.u.; hence, the range of the distribution, which is centred at z_w=100 m.u., is from z=20 up to z=180. Note that for a uniform distribution z_w=z_n

$$f_{\mathrm{w}}(z)=\frac{1}{w} \quad \text{for} \quad z_{\mathrm{w}}-\tfrac{1}{2}w<z<z_{\mathrm{w}}+\tfrac{1}{2}w \tag{97a}$$

$$f_{\mathrm{w}}(z)=0 \quad \text{else} \tag{97b}$$

for the calculation of the effect of the width of the distribution. Figure 49 shows the results of the strength as a function of the relative width, wz_{w}^{-1}, for the rod diameter $2r$=10 nm. For example, for the point in this figure positioned on the line z_{w}=100 m.u. with an "x" coordinate of $12.5wz_{\mathrm{w}}^{-1}=5$, the width equals w=40 m.u. yielding the distribution range $80<z<120$ and the fibre strength is 2.69 GPa. For the point on the line z_{w}=100 m.u. with an "x" coordinate of $12.5wz_{\mathrm{w}}^{-1}=15$, the width equals w=120 m.u. yielding the distribution range $40<z<160$ and the fibre strength is 2.52 GPa. A relatively small decrease in strength is calculated with increasing width of the distribution, while the average molecular weight is kept constant. Apparently, the detrimental effect on the strength due to an increase of the fraction of short chains in wider distributions is almost compensated by the simultaneous increase of the fraction of long chains. However, for smaller rod diameters (not shown) the effect becomes less pronounced and almost disappears for $2r$=0.5 nm.

3.3 Effect of Low Molecular Weight Fraction on Fibre Strength

All computations so far have been performed with the full width of the chain length distributions, i.e. chains with aspect ratios $L/d<10$ have also been included. However, the effects of stress transfer across chain ends and the stress concentrations may become important below this aspect ratio. In the theory by Yoon these effects are neglected. In particular, for the Flory distribution containing a relatively large proportion of very short chains, the effects may be considerable. Therefore, calculations are performed in which the negative effect of the very short rods is approximated by assuming that for an aspect ratio $L/d\leq 4$ the contribution of these rods to the strength is set to zero.

The Flory distribution depicted in Fig. 44 has a large fraction of very short chains, which suggests that elimination of this part of the chain length distribution may have a strong positive effect on the fibre strength. However, there are two reasons why the effect of very short chains on the strength is very small indeed. Firstly, the maximum stress at the centre of the chain in short chains is much smaller than in long chains; thus, short chains contribute relatively less to the fibre strength than long chains. Secondly, the important distribution for the determination of the strength is the crossing length distribution, which is the number of chains of each length crossing a given fibre cross section. This distribution is given by the function

$$h(z)=\frac{zf(z)}{z_{\mathrm{n}}} \tag{98}$$

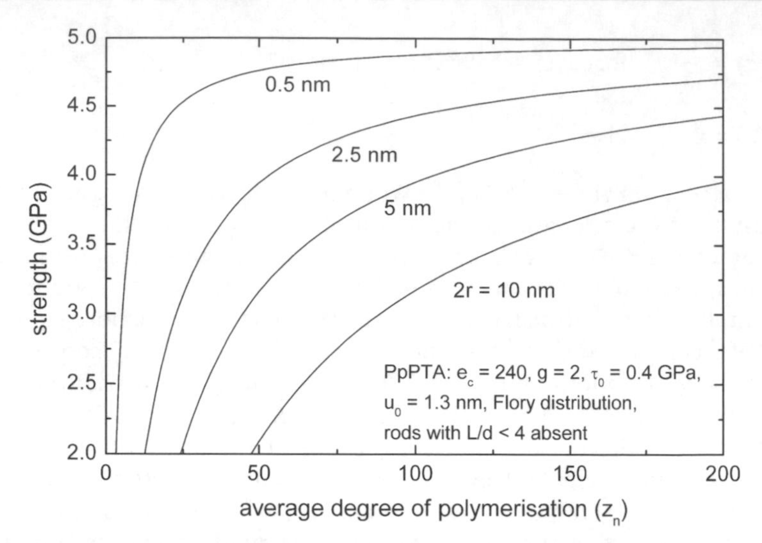

Fig. 50 The ultimate strength for a Flory distribution in which rods with an aspect ratio $L/d \leq 4$ do not contribute to the strength

By means of Eq. 98 it is expressed that long chains have a larger probability to cross a fibre cross section. In the paper by Yoon no distinction is made between the chain length distribution $f(z)$ and the crossing length distribution $h(z)$. In the latter distribution, short chains have little effect on the strength. As an example the strength has been calculated for a Flory distribution in which all rods with an aspect ratio $L/d \leq 4$ have been eliminated, but they were still included in the calculation of the average degree of polymerisation. Comparison of Fig. 50 with Fig. 45 shows that only for the largest diameter $2r=10$ nm is there a slight difference for small values of z_n.

The model is applied to calculate the effect of the low molecular weight fraction on the strength using the experimental molecular weight distributions of PpPTA polymer samples. Table 1 presents the experimental data of the key parameters of three samples of PpPTA polymer with different distributions, while Fig. 51 presents the experimental distribution of one of the samples. With a computer program that is based on the model by Yoon and that takes account of the crossing length distribution, the number-average molecular weight and the ultimate fibre strength were calculated, using the molecular weight distributions derived from the experimental distributions. In a similar way the average molecular weight and ultimate fibre strength have been calculated for each of the three distributions after the low molecular fraction was eliminated. The distributions were cut off at $M_w \leq 6{,}200$ or for a DP$\leq$26. Table 2 shows that elimination of the low molecular weight fraction yields a considerable increase of the number-average molecular weight (z_n). The increase of the ultimate strength by this elimination is less dramatic and strongly depends on the aspect ratio of the elemental building block. In the case where the chain itself is the elemental building block, the

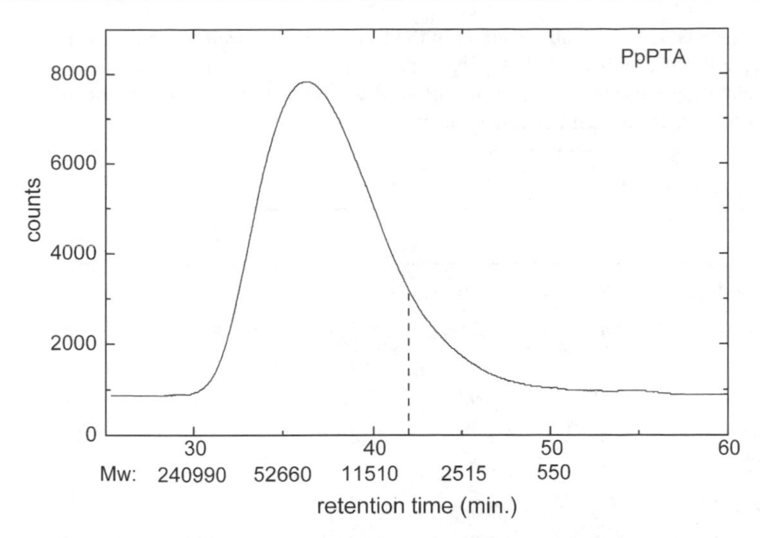

Fig. 51 The experimental curve of the molecular weight distribution of a PpPTA polymer dissolved in sulphuric acid determined with size-exclusion chromatography. The relation between molecular weight and retention time t (in minutes): $^{10}\log(M_w)=0.345-0.1321\ t$. The position of the *vertical line* at t=42 min corresponds with M_w=6,200

Table 1 Key parameters of the molecular weight distribution for three samples of PpPTA polymer from which yarns have been spun. M_z is the z-average molecular weight [52]

Sample	M_n	M_w	M_z	% Material $M_w\leq 6{,}200$
QCX 9906-06	11,500	33,400	55,000	12.1
QCX 9903-09	13,800	42,200	68,000	8.9
QCX 9905-16	15,200	45,700	72,000	7.8

effect of the elimination of the low molecular weight fraction is rather small, viz. for z_n=48 the strength increases from 4.76 to 4.88 GPa or 2.5%. In the case of fibrils with a diameter of 5 nm and a length equal to 59 monomers or 77 nm the effect is more substantial, viz. 7.6%. For large rod diameters some caution is required as the model holds only for aspect ratios $u_a/(2r)\geq 10$.

A few experimental results on the relation between strength and the molecular weight have been reported. Figure 52 shows for low viscosity values a clear relation between inherent viscosity and strength of PpPTA yarns [53]. In Table 2 the ultimate fibre strength is given on the basis of Yoon's model for the three PpPTA polymer samples listed in Table 1, using the observed distributions of these polymers measured by size-exclusion chromatography. Comparison of the differences between the calculated strength as a function of the chain length in Table 3 with the observed strength indicates that the chain is the elemental building block.

Table 2 Results of the calculated effect of the low molecular weight fraction on the ultimate fibre strength for the three PpPTA polymer samples listed in Table 1. The constants used in the calculation are e_c=240, g=2, τ_0=0.4 GPa and u_0=1.3 nm. The maximum ultimate strength for infinite long chains using these constants is 5 GPa

Sample	z_n	Strength (GPa) versus rod diameter $2r$ (nm)			
		10	5	2.5	0.5
QCX 9906-06	48	2.59	3.42	4.04	4.76
QCX 9906-06 $M_w \geq 6{,}200$	96	2.91	3.80	4.38	4.88
QCX 9903-09	59	2.92	3.68	4.22	4.81
QCX 9903-09 $M_w \geq 6{,}200$	112	3.17	3.96	4.47	4.89
QCX 9905-16	64	3.03	3.77	4.28	4.82
QCX 9905-16 $M_w \geq 6{,}200$	119	3.26	4.02	4.50	4.90

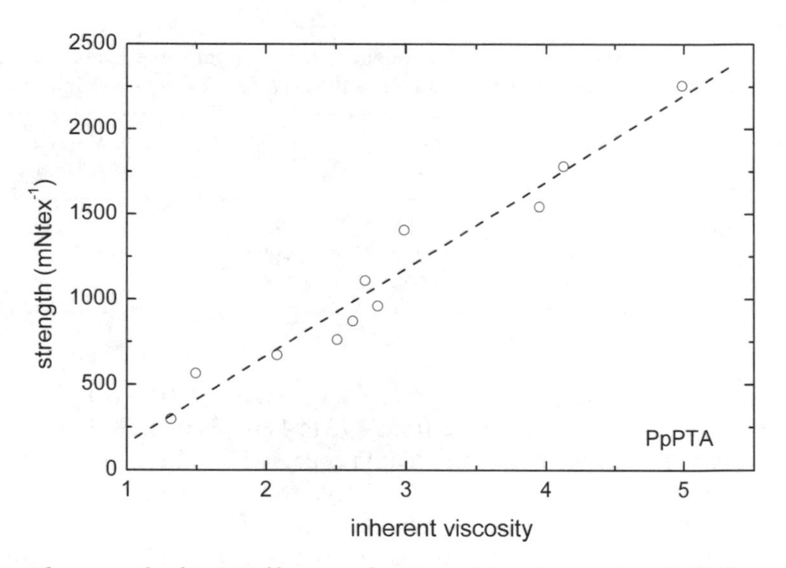

Fig. 52 The strength of PpPTA fibres as a function of the inherent viscosity [53]

Table 3 Comparison of the ultimate strength σ_L calculated with Yoon's model with the observed fibre strength (see Table 2)

Sample/fibre	z_n	Calculated strength (GPa) versus rod diameter 2 r (nm)				Observed strength (GPa)
		10	5	2.5	0.5	
QCX 9906-06	48	2.59	3.42	4.04	4.76	3.60
QCX 9903-09	59	2.92	3.68	4.22	4.81	3.66
QCX 9905-16	64	3.03	3.77	4.28	4.82	3.81

3.4 Theoretical Effect of Molecular Weight on the Modulus

In the preceding section we discussed the effect of the molecular weight on the ultimate strength. There may also be another explanation for the increase in strength with increasing molecular weight. Increasing the average chain length may improve the degree of orientation in the domains of the anisotropic solution, which results for the same stretch ratio in a higher modulus leading to an increase in fibre strength. To estimate the effect of the molecular weight on the orientation, modulus and strength of PpPTA fibres, the model by Picken is applied [47–51]. The modulus is related to the degree of orientation by:

$$\frac{1}{E} = \frac{1}{e_c} + \frac{\langle \sin^2\Theta \rangle_E}{2g} \tag{99}$$

The orientation in the model is described in terms of the order parameter $\langle P_2 \rangle$, so the relation between $\langle P_2 \rangle$ and $\langle \sin^2\Theta \rangle$ is required:

$$\langle \sin^2\Theta \rangle = \tfrac{2}{3}(1 - \langle P_2 \rangle) \tag{100}$$

Note that only for a high degree of orientational order, as is normally the case for spinning conditions, can the approximation $\langle \sin^2\Theta \rangle_E \approx \langle \sin^2\Theta \rangle$ be used. Also, in the case of high orientational order the $\langle P_2 \rangle$ value is given to a good approximation by

$$\langle P_2 \rangle = 1 - 0.22 \left(\frac{T}{T_{ni}} \right)^3 \tag{101}$$

where T is the temperature at which the orientational order is frozen in. Therefore in this model, the effect of molecular weight can only occur via the clearing temperature T_{ni}. To estimate the effect of M_w on T_{ni} the following relation is used

$$k_B T_{ni} = 0.22 \cdot \zeta^* c^2 L_d^2 (T_{ni}) \tag{102}$$

where k_B is the Boltzmann constant and ζ^* is a constant determining the absolute temperature scale. As shown by Eq. 102, in the extension of the Maier–Saupe model to polymer solutions the strength of the potential contains the polymer concentration c (which is constant in the present case) and the contour projection length $L_d(T)$. The contour projection length is the average length of the projection of the worm-like chain along the direction of the first segment and is given by

$$L_d(T) = L_P \frac{1 - \exp\left(-\frac{L_C T}{L_P T_P}\right)}{T/T_P} \tag{103}$$

Here, L_C is the contour length, L_P is the persistence length and T_P the temperature at which L_P is measured. The contour projection length determines whether we are dealing with rod-like polymer chains (for low M_w or $L_C<L_P$) or with Gaussian coils (for high M_w, $L_C>L_P$). In the first case $L_d(T)$ is nearly independent of T and is just the end-to-end length of the rod, whereas in the second case $L_d(T)\approx L_P T_P/T$, which is just the persistence length at temperature T (L_P=29 nm at T_P=20 °C=293 K). Therefore, to estimate whether the molecular weight influences the degree of orientation via a change of T_{ni} we only have to investigate if there is a difference in the value of $L_d(T_{ni})$. It can now easily be seen that such an effect for PpPTA is not to be expected. If for instance we take M_w=35,000, then the contour length $L_C=M_w u_0/(\text{mole})$=35,000·12.8/238=188 nm. The clearing temperature of a 20% (w/w) solution will be about 290 °C (560 K) using the expression $T_{ni}=76\cdot c^{0.66}$, and we will take the spinning bath temper-

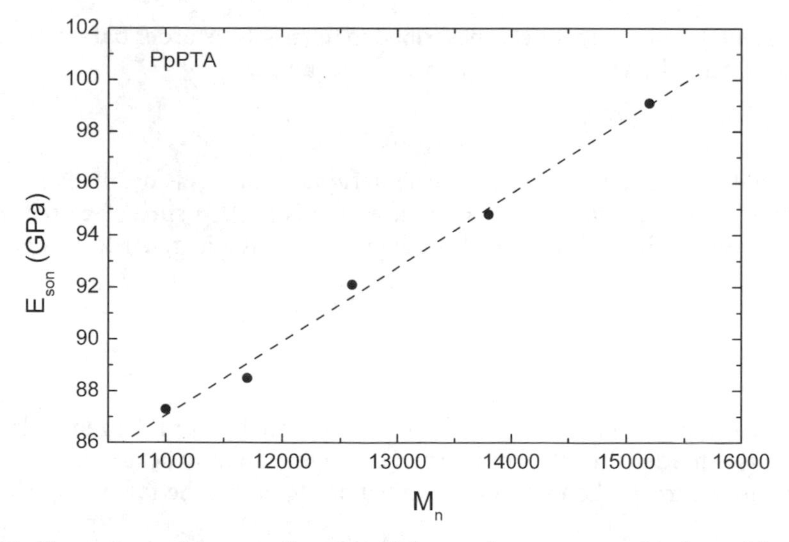

Fig. 53 The sonic modulus as a function of the number-average molecular weight for a series of experimental PpPTA yarns with different molecular weights specified in Table 4

Table 4 The molecular weight, sonic modulus and strength of experimental PpPTA yarns. The molecular weight of a single monomer is 238

Fibre sample	M_n	M_w	M_z	E_{son}(GPa)	σ_b(GPa)
QCX 9903-06	11,000	34,700	58,000	87.3	3.60
QCX 9903-07	11,700	36,900	61,000	88.5	3.71
QCX 9903-08	12,600	39,200	64,000	92.1	3.59
QCX 9903-09	13,800	42,200	68,000	94.8	3.66
QCX 9905-16	15,200	45,700	72,000	99.1	3.81
QCX 9906-06	11,500	33,400	55,000	98.1	3.58

ature as T=20 °C. Filling this into the expression for $L_d(T)$ shows that the exponential term is in the order of $6 \cdot 10^{-6}$, so that there is a negligible effect on the clearing temperature for this range of molecular weight. The molecular weight will only start to have a noticeable influence when the argument of the exponential term is about –3 or when L_C is about $1.5L_P$ or less (e^{-3}=0.05, $T_{ni}T_P^{-1} \approx 1.9$), which occurs for $M_w \leq$8,400 amu or $\eta_{rel.} \leq 1.7$. In the above arguments we have avoided calculating T_{ni} consistently from Eq. 102 as this only complicates matters without yielding any further insight. The conclusion of this exercise is that the molecular weight is not expected to have any effect on the orientation for the present M_w range of 35,000–45,000 via a change in the orientational order of the spinning solution. The remaining mechanism for influencing the modulus with the molecular weight would be via the rate of elongational flow in the drawing process. However, in the case of high alignment at high polymer concentrations the additional effect of flow can be neglected. This is confirmed by the experimental observation that there is little or no effect of the absolute spinning rate on the fibre modulus.

Spinning experiments, however, present a different picture as shown in Table 4, which lists the results for PpPTA yarns with different molecular weights. Surprisingly, as shown in Fig. 53, there is a very strong correlation between the sonic modulus and M_n. Furthermore, there is a positive correlation between strength and M_n, and hence also between strength and modulus. Therefore, it seems likely that the increase in strength is caused by the increase in modulus as predicted by the model discussed in Sect. 2. Probably the degree of molecular alignment may slightly increase with increasing molecular weight, due to the increased elasticity of the solution and the applied elongational flow in the air gap of the spinning process of liquid crystalline solutions [51]. The observed change in modulus (from 87 to 99 GPa) indicates that the change in the $\langle P_2 \rangle$ values is about 0.9618–0.9544=0.0074. This is, of course, only a minor change, which is why these effects were not included in the model, but it is clearly an important factor in spinning technology where a small rheological change can have a large effect.

3.5 Conclusions

The few observations seem to indicate that the dependence of the fibre strength on the chain length is stronger than the dependence of the ultimate strength on the chain length as calculated with the molecular composite model. The calculations for different kinds of chain length distributions show that for a rod diameter equal to the chain diameter, and for a perfectly parallel-oriented fibre, an increase of the ultimate strength of only 4% is obtained for a DP increase from 50 to 200. Hence, an increase of M_n ($=z_n$) from 48 to 65 as shown in Table 3 should hardly give any increase of the strength in PpPTA fibres. Yet, the observed strength increases from 3.6 to 3.81 GPa or by 5.8%. Therefore, it appears that the "translation" of macrocomposite parameters to molecular composite parameters is not without difficulties. Except for the rather complicated model of Termonia, Smith and Meakin, there is presently no other model that gives an estimate of the chain length dependency on the fibre strength [12, 13].

The calculation of the strength for various aspect ratios shows that the fibre strength should benefit from a finer morphology. As shown in Sect. 5 this can be achieved, for example, by spinning smaller filament counts. When the observed increase of the fibre strength is larger than the value estimated on the basis of the increase of DP, the model also suggests that the diameter of the rod or the building block in a real fibre might be considerably larger than the chain diameter. In this case the function of the rod is probably taken over by fibrils. In more general terms, a considerable change in molecular weight may result in a change of certain micromorphological aspects that can influence the fibre

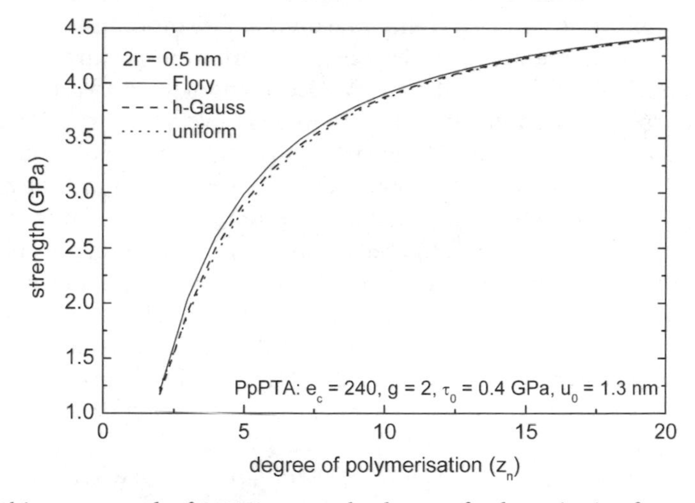

Fig. 54 Ultimate strength of PpPTA versus the degree of polymerisation for a rod diameter of 0.5 nm. Comparison of the results calculated for the Flory, the half-Gauss and the uniform distributions

strength. For example, the chain length distribution might also influence the morphology during coagulation. As longer chains have lower diffusion velocities, the cross section of the aggregates formed by long chains during coagulation may be smaller than for shorter chains, resulting in a finer morphology with higher aspect ratios. This argument also suggests that a smaller filament diameter may lead to a finer morphology. Of course, in this discussion it is assumed that the increase in molecular weight will not alter the modulus of the fibre.

The model demonstrates that very short chains have little effect on the fibre strength. As shown by Fig. 54, the effect of the shape of the chain length distribution for a rod diameter of 0.5 nm (equal to the chain diameter) is relatively small. For much thicker rods, $2r$=10 nm, Fig. 55 shows that the differences between the various distributions extend to high degrees of polymerisation. Because the effect of very short chains on the fibre strength is found to be very small, the result that the Flory distribution yields the highest strength is solely due to the very long chains in this distribution. The calculation shows that a clear effect of the width of the distribution is only found for large rod diameters or small aspect ratios.

By applying the model to experimental molecular weight distributions we have shown that elimination of the low molecular weight fraction yields a considerable increase of the DP. The calculated increase in strength is rather small and strongly depends on the aspect ratio of the elemental building block. Finally, it cannot be ruled out that the improvement of the fibre strength by the increase of the average chain length might be explained by a slight change of the rheological behaviour of the solution during spinning.

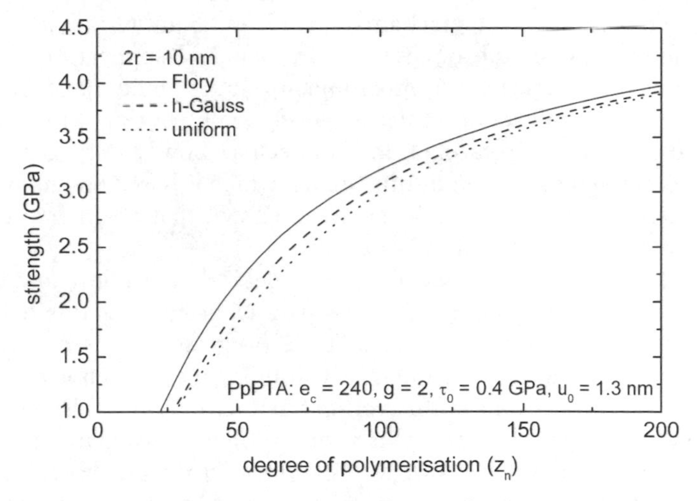

Fig. 55 Ultimate strength of PpPTA versus the degree of polymerisation for a rod diameter of 10 nm. Comparison of the results calculated for the Flory, the half-Gauss and the uniform distributions

4
Dependence of the Fibre Strength on the Time and the Temperature

4.1
Introduction

In the previous sections the theoretical relations describing the strength of a polymer fibre as a function of the intrinsic parameters, such as the chain modulus, the modulus for shear between adjacent chains, the orientation distribution and the chain length distribution, have been discussed. In this section the dependence of the strength on the time and the temperature will be investigated.

According to the continuous chain model the elongation of a polymer fibre is brought about by the elastic extension of the chain determined by the covalent bonds, and by the viscoelastic shear deformation of the domain being the building block of the fibre. This shear deformation is governed by the secondary bonds between the chains [1–10]. Due the nature of this bonding the strength of a polymer fibre is a function of the test speed or time and of the temperature. In a creep failure experiment a fibre is loaded to a constant stress σ_b and the time-to-failure t_b is measured. The observed lifetime relation for many polymer fibres is

$$\log(t_b) = C_1 - C_2\sigma_b \tag{104}$$

in which the parameters C_1 and C_2 are a function of temperature. Because in a creep failure experiment the creep stress is lower than the fibre strength determined in a normal tensile test, this relation cannot be explained by a failure criterion based on a constant critical shear stress.

An understanding of the mechanism of creep failure of polymer fibres is required for the prediction of lifetimes in technical applications. Coleman has formulated a model yielding a relationship similar to Eq. 104. It is based on the theory of absolute reaction rates as developed by Eyring, which has been applied to a rupture process of intermolecular bonds [54]. Zhurkov has formulated a different version of this theory, which is based on chain fracture [55]. In the preceding sections it has been shown that chain fracture is an unlikely cause for breakage of polymer fibres.

The structural picture of the Coleman model is that in a fibre under a load, the slippage of adjacent and more or less parallel oriented chains reduces the stress transfer between the chains as time progresses. This creep rupture process continues as long as the interchain bonding is strong enough to sustain the load. At a critical or minimum overlap the fracture process is initiated and results in the failure of the fibre. This assumption is equivalent to stating that there is a critical amount of distortion that the microscopic structure can tolerate before the bonding or the activation energy U rapidly decreases to zero and a catastrophic breakdown occurs. Coleman found evidence for this hypothesis in the observation that the elongation at break ε_b is far less sensitive to the rate of loading than the tensile strength. In addition, the hypothesis is also supported by the observation that a short loading at a stress slightly below the tensile strength of a polymer fibre does not lower the strength.

In view of the development of the continuous chain model for the tensile deformation of polymer fibres, we consider the assumptions on which the Coleman model is based as too simple. For example, we have shown that the resolved shear stress governs the tensile deformation of the fibre, and that the initial orientation distribution of the chains is the most important structural characteristic determining the tensile extension below the glass transition temperature. These elements have to be incorporated in a new model.

As shown in Sect. 2, the fracture envelope of polymer fibres can be explained not only by assuming a critical shear stress as a failure criterion, but also by a critical shear strain. In this section, a simple model for the creep failure is presented that is based on the logarithmic creep curve and on a critical shear strain as the failure criterion. In order to investigate the temperature dependence of the strength, a kinetic model for the formation and rupture of secondary bonds during the extension of the fibre is proposed. This so-called Eyring reduced time (ERT) model yields a relationship between the strength and the load rate as well as an improved lifetime equation.

4.2 Simple Derivation of the Lifetime of a Polymer Fibre under Constant Load

The resolved shear stress $|\tau|=\sigma\sin\theta\cos\theta$ brings about a relative displacement of adjacent chains as shown in Figs. 56 and 57. As has been argued in Sect. 2.1, shear failure is a much more likely cause for fibre fracture than chain scission. This failure mode is in agreement with the observation that fibres having no melting temperature show a fibrillar fracture morphology. However, the fracture criterion involving shear failure does not exclude the possibility of chain scission, which may occur in the final stages of the fracture process.

Before we enter upon the discussion of the lifetime of a polymer fibre, a brief presentation is given of the creep theory of polymer fibres according to the continuous chain model [7–10]. The total fibre strain is given by the sum of the

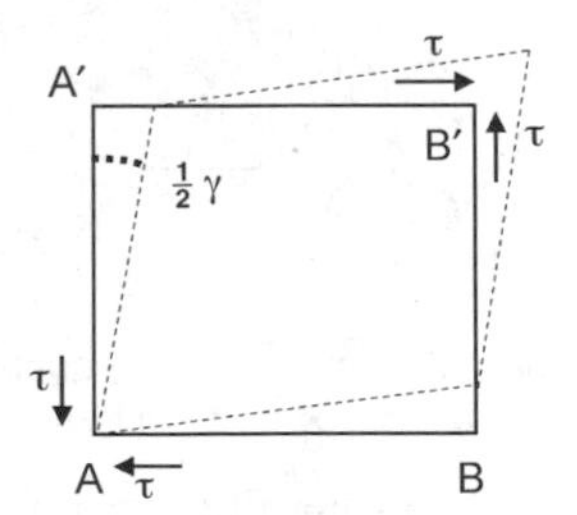

Fig. 56 The effect of the four shear stresses on the shape of the domain AA′BB′. In the unloaded condition the chain direction is parallel to AB. The shear angle equals $\varepsilon_{13}=1/2\gamma$

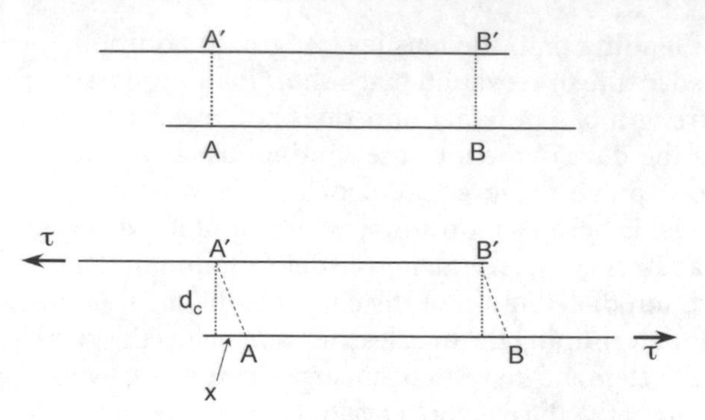

Fig. 57 Due to the shear deformation of the domain, the relative displacement of two adjacent chains equals $x=d_c\tan\gamma$, d_c being the distance between the chains

contributions from the chain extension and from the chain rotation due to the shear deformation of the domain

$$\varepsilon_f = \frac{\sigma\langle\cos^2\theta\rangle}{e_c} + \frac{\langle\cos\theta\rangle - \langle\cos\Theta\rangle}{\cos\Theta} \tag{7a}$$

where Θ is the initial orientation angle of the chain axis at zero load, θ the angle at a stress σ and e_c the chain modulus. The viscoelastic properties of a polymer fibre are well described by the schematic rheological model depicted in Fig. 58. It consists of a series arrangement of a "tensile" spring with pure elastic properties representing the chain modulus, and a "shear" spring with viscoelastic properties characterising the shear deformation of the domain. The fibre modulus is given by

$$\frac{1}{E(t)} = \frac{1}{e_c} + \frac{\langle\sin^2\theta(t)\rangle_E}{2g(t)} \tag{105}$$

Because of the viscoelastic nature of the polymer fibre the modulus is a function of the rate of measurement, which is indicated by $E(t)$. We might as well

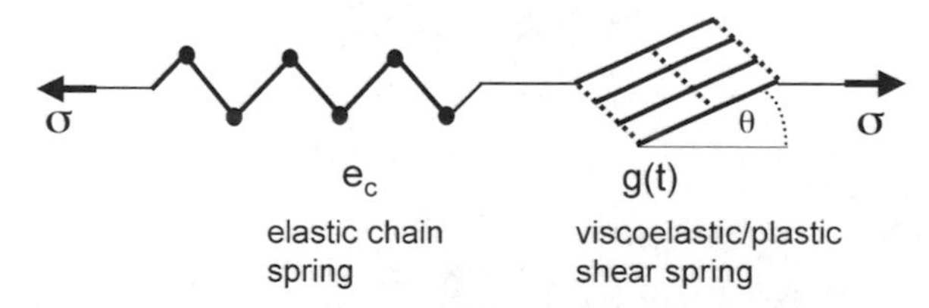

Fig. 58 The rheological model of a polymer fibre consists of a series arrangement of an elastic "tensile" spring representing the chain modulus, e_c, and a "shear" spring, $g(t)$, with viscoelastic and plastic properties representing the intermolecular bonding

have written $E(\nu)$, as the time dependency also implies a frequency dependency. Whereas the covalent bonds provide the chain with purely elastic properties, the viscoelasticity of polymer fibres has its origin in the intermolecular bonding indicated by the time- or frequency-dependent function of the shear modulus $g(t)$ or $g(\nu)$.

During the creep of PET and PpPTA fibres it has been observed that the sonic compliance decreases linearly with the creep strain, implying that the orientation distribution contracts [56, 57]. Thus, the rotation of the chain axes during creep is caused by viscoelastic shear deformation. Hence, for a creep stress larger than the yield stress, σ_y, the orientation angle is a decreasing function of the time. Consequently, we can write for the viscoelastic extension of the fibre

$$\varepsilon_f(t) = \frac{\sigma\langle\cos^2\theta(t)\rangle}{e_c} + \frac{\langle\cos\theta(t)\rangle - \langle\cos\Theta\rangle}{\langle\cos\Theta\rangle} \tag{106}$$

In order to simplify the discussion and keep the derivation of the formulae tractable, a fibre with a single orientation angle Θ is considered. In a creep experiment the tensile deformation of the fibre is composed of an immediate elastic and a time-dependent elastic extension of the chain by the normal stress $\sigma\cos^2\theta(t)$, represented by the first term in the equation, and of an immediate elastic, viscoelastic and plastic shear deformation of the domain by the shear stress, $|\tau|=\sigma\sin\theta(t)\cos\theta(t)$, represented by the second term in Eq. 106.

The total contribution of the shear strain to the fibre strain is the sum of the purely or immediate elastic contribution involving the change in angle, $\Delta\theta_e=\theta_0-\Theta$, occurring immediately upon loading of the fibre at $t=0$, and the time-dependent or viscoelastic and plastic contribution $\Delta\theta(t)=\theta(t)-\theta_0$ [7–10]. According to the continuous chain model for the extension of polymer fibres, the time-dependent shear strain during creep can be written as

$$\varepsilon_{13}(t) = \tan(\theta(t) - \Theta) \approx \frac{\tau(t)}{2g} + \frac{j(t)\tau(\theta_0)}{2} \tag{107}$$

where $\tau(\theta_o)=-\sigma\sin\theta_0\cos\theta_0$ is the shear stress at $t=0$, $\tau(t)=-\sigma\sin\theta(t)\cos\theta(t)$ and $j(t)$ is the creep compliance. The second term in Eq. 107 yields the viscoelastic contribution to the fibre strain

$$\varepsilon_f^v(t) = \tfrac{1}{2}j(t)\tau_n \tag{108}$$

where τ_n is the normalised shear stress given by

$$\tau_n = 2g\left(\frac{1}{E} - \frac{1}{e_c}\right)\frac{\sigma}{\left(1 + \frac{\sigma}{2g}\right)^3} \tag{109}$$

Figure 59 shows that the creep rate calculated as a function of the modulus using Eqs. 108 and 109 agrees well with the experimental data of a series of PpPTA

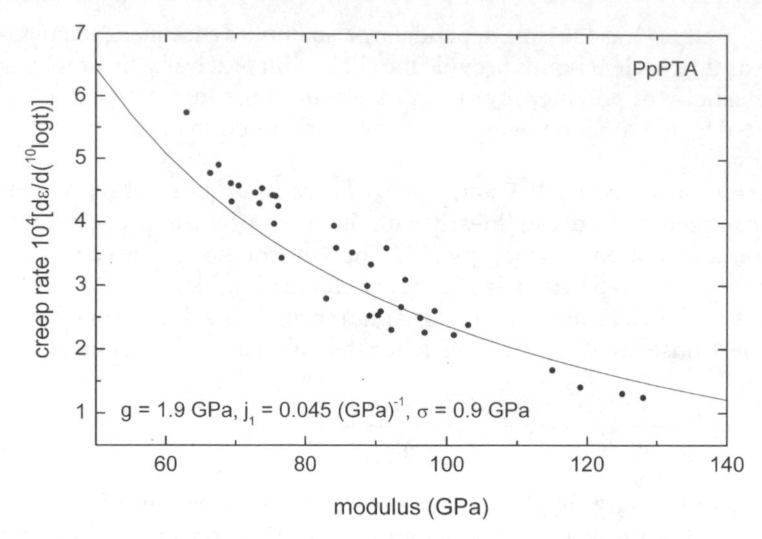

Fig. 59 Comparison of the calculated creep rate with the observed data of a series PpPTA yarns with different moduli for a creep stress of 0.9 GPa

yarns. This creep model has been confirmed for PpPTA fibres up to a stress of 2 GPa [7].

It can now be proposed that a maximum shear strain is a good fracture criterion for polymer fibres. As a result of the increasing shear between adjacent chains, the intermolecular bonds are distorted to such a degree that the bonding becomes insufficient for the stress transfer between the chains. When a distortion is reached at which the overlap is too small to sustain the load, we assume that the fracture of the fibre is initiated. It is important to realise that this fracture criterion of a critical shear strain does not involve large parallel displacements of adjacent chains. Kooijman and Kroon-Batenburg have calculated that in the case of two hydrogen-bonded chains in a perfect PpPTA crystal, only a relatively small displacement of about 0.17 nm or 0.3 radian is necessary to reach a point where the attracting force starts to decrease [38, 39]. As demonstrated in Sect. 2, the hyperbolic shape of the fracture envelope can also be explained by this criterion. We now postulate that during tensile deformation of a polymer fibre the sum of the absolute value of the elastic shear strain, $|\Delta\theta_e|$, and the absolute value of the viscoelastic and plastic shear strain, $|\Delta\theta(t)|$, cannot exceed a particular value β, otherwise the fibre will break. This criterion resembles to some degree the critical distortion concept by Coleman, the difference being that Coleman used a maximum fibre strain instead of a maximum shear displacement of the chains [54]. The criterion of the maximum rotation or maximum shear strain can therefore be written as

$$(\Theta - \theta_0) + |\Delta\theta(t_b)| = \beta \quad \text{or} \quad \Theta - \theta_b = \beta \tag{110}$$

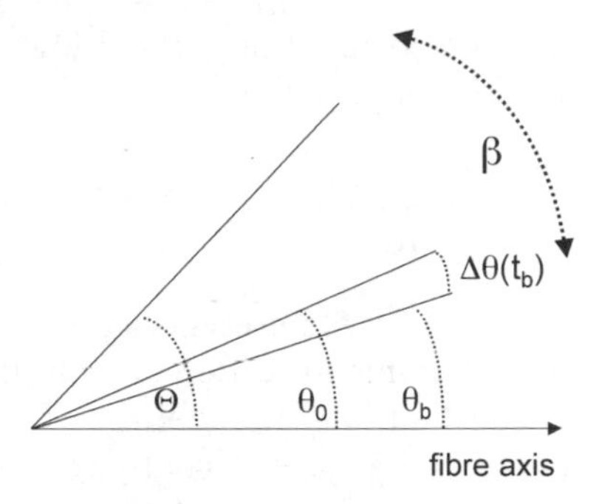

Fig. 60 The angles of the domain axis in the unloaded state, Θ: after loading at t=0, θ_0; at the time of fracture t_b, θ_b; and the maximum shear strain β

where t_b is the time-to-failure of the fibre, see also Fig. 60. (Note that in a creep and in a tensile experiment the rotation of the chain axis towards the fibre axis, $\Delta\theta$, is defined as a negative quantity).

The maximum shear strain in engineering units is γ_b=2β, which corresponds to the value $p(4d_c)^{-1}$ in the Frenkel model discussed in Sect. 2.5.1. The relative displacement, x, of adjacent chains shown in Fig. 57 is limited by the maximum value 2β; for larger displacements the attracting force decreases rapidly and failure is initiated. It is further assumed that $|\Delta\theta(t_b)| \ll \Theta-\theta_0$. Describing the creep with Eq. 107 the condition for fracture becomes

$$\beta \approx \frac{|\tau(t_b)|}{2g} + \frac{j(t_b)|\tau(\theta_0)|}{2} \tag{111}$$

Because $\tau(t) \cong \tau(\theta_0) - \sigma\Delta\theta(t)$ and $|\Delta\theta(t_b)| = \beta - (\Theta - \theta_0) = \beta - |\tau(\theta_0)|(2g)^{-1}$, the shear stress at fracture can be written as

$$\tau(t_b) = \tau(\theta_0) - \sigma[\beta - \tau(\theta_0)(2g)^{-1}] \tag{112}$$

In order to calculate θ_0, we use for the solution of the equation for the shear strain

$$\tan(\theta_0 - \Theta) = -\frac{\sigma}{2g}\sin\theta_0\cos\theta_0 \tag{113}$$

the approximate analytical expression:

$$\tan\theta_0 = \frac{\tan\Theta}{\left(1 + \dfrac{\sigma}{2g}\right)} \tag{114}$$

By adopting a logarithmic creep function $j(t)=j_1{}^{10}\log(t)$, Eq. 111 yields the following equation for the lifetime of a fibre

$$^{10}\log(t_b) = \frac{\beta(2g+\sigma_b)^2}{2j_1 g^2 \sigma_b \tan\Theta} - \frac{(2g+\sigma_b)}{2j_1 g^2} \tag{115}$$

For $\sigma_b \to 0$ the lifetime $t_b \to \infty$, so Eq. 115 presents a non-linear relation between $\log(t_b)$ and the creep stress σ_b, which is different from the Coleman relation. According to Eq. 115, at constant load the lifetime of a fibre decreases with increasing orientation parameter. Figure 61 compares the observed data for a PpPTA fibre by Wu et al. with the calculated lifetime curve using the parameter values β=0.08, tanΘ=0.1483, g=1.6 GPa, j_1=0.032 (GPa)$^{-1}$, which implies a fibre with a sonic modulus of 91.8 GPa [30]. As shown by Wu et al., fibres that were tested at high stresses had shorter lifetimes than those calculated from the experimental lifetime relation.

For t_b=1 s, Eq. 115 yields an expression for the strength based only on the shear failure of the domain

$$\sigma_b^s = \frac{2\beta g}{\tan\Theta - \beta} \qquad \text{for} \quad \Theta > \arctan\beta \tag{116}$$

Equation 116 was also derived in Sect. 2. It shows that the fibre strength according to the shear failure criterion increases with improved alignment of the chains, and that it is proportional to the shear modulus g.

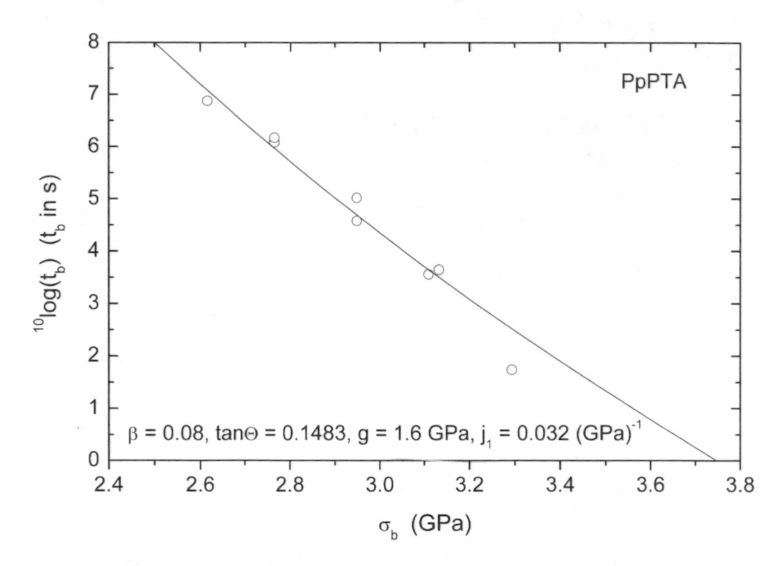

Fig. 61 The calculated lifetime curve according to Eq. 115 compared with the experimental results on PpPTA filaments by Wu [30]

From Eqs. 115 and 116 the relative strength change per decade at t=1 s is calculated

$$\left(\frac{\Delta\sigma_b}{\sigma_b}\right)_{\text{decade}} = -j_1 g \tag{117}$$

Creep measurements of PpPTA show that $0.025<j_1$ $(\text{GPa})^{-1}<0.05$, thus for g=1.6 GPa the expected range of the strength loss is $0.04<j_1<0.08$. The observed value for PpPTA fibres is about 5%.

The maximum shear strain criterion is now applied for the calculation of the creep curve up to fracture for increasing creep stress. The total creep strain of the fibre, $\varepsilon_f(t)$, is the sum of the elastic strain, ε_f^e, and the viscoelastic plus plastic strain, $\varepsilon_f^v(t)$,

$$\varepsilon_f(t) = \varepsilon_f^e + \varepsilon_f^v(t) \tag{118}$$

Since for a well-oriented fibre $\varepsilon_f^e \approx \sigma/E$, the total creep strain is given by

$$\varepsilon_f(t) \approx \frac{\sigma}{E} + \tfrac{1}{2} j(t)\tau_n \tag{119}$$

Assuming logarithmic creep, $j(t)=j_1{}^{10}\log(t)$, Eqs. 118 and 119 allow the calculation of the creep curves up to fracture. The results are depicted in Fig. 62. Note that for increasing creep stress the slope of these curves decreases.

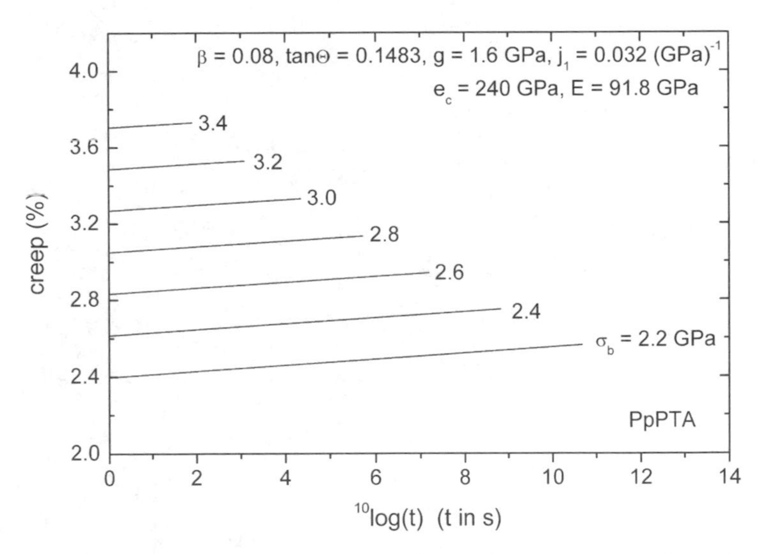

Fig. 62 The creep failure curves of a PpPTA fibre with an initial modulus of 91.8 GPa calculated with the same parameters as used in Fig. 61

4.3
Relationship Between Strength and Load Rate Derived from the Eyring Reduced Time Model

For the investigation of the time and the temperature dependence of the fibre strength it is necessary to have a theoretical description of the viscoelastic tensile behaviour of polymer fibres. Baltussen has shown that the yielding phenomenon, the viscoelastic and the plastic creep of a polymer fibre, can be described by the Eyring reduced time (ERT) model [10]. The shear deformation of a domain brings about a mutual displacement of adjacent chains, the

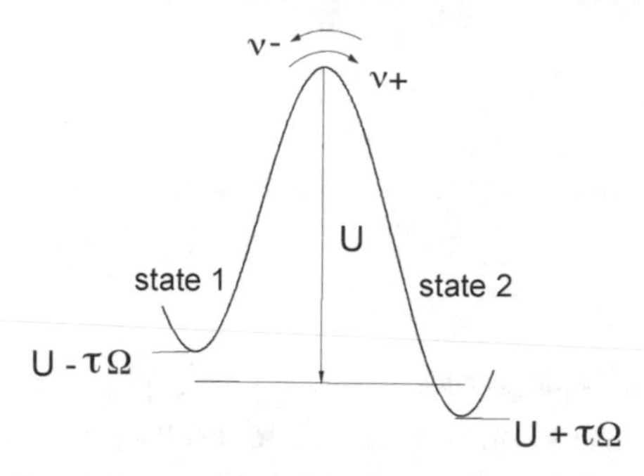

Fig. 63 The Eyring reduced time model involves the activated site model for plastic and viscoelastic shear deformation of adjacent chains

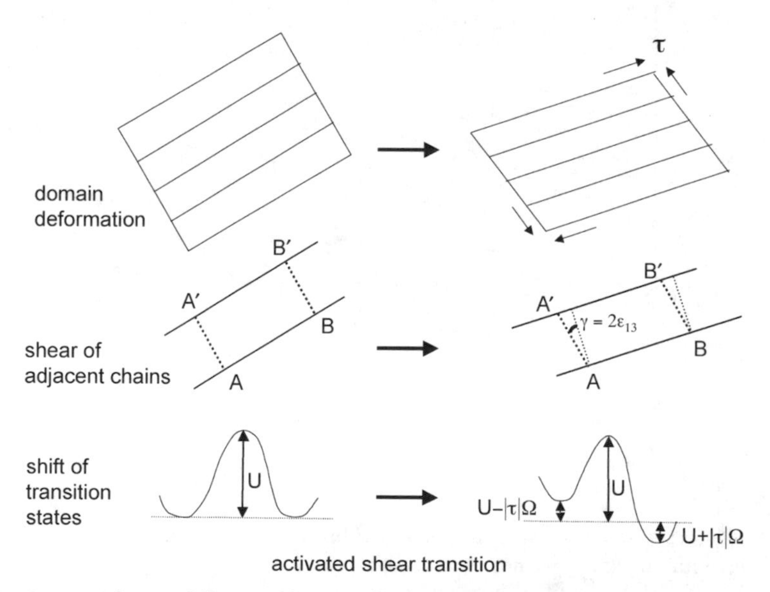

Fig. 64 Creep and creep failure can be modelled by the time-dependent shear deformation as described by the Eyring reduced time model

amount of which is determined by the secondary bonding between the chains. The straining of this bonding is now modelled as an activated shear transition between two states separated by an energy barrier U, as depicted in Figs. 63 and 64. In a perfect crystal there is only a single value for U, but in semi-crystalline and para-crystalline fibres a distribution $I(U)$ can be expected which is associated with the energies of the various conformations of the secondary bonds between the chains. Hence, the function $I(U)$ is a material property and, similar to the elastic constants e_c and g, governs the elastic and viscoelastic tensile deformation of the fibre.

In Fig. 63 the occupation of state 1 is equal to N_1, the occupation of state 2 is equal to N_2 and the total occupation is equal to $N_1+N_2=N$. The viscoelastic and plastic shear strain is proportional to the decrease of the occupation of state 1 or proportional to the increase of the occupation of state 2. Without external stress the probability for transition from state 1 to state 2 (v^+) is proportional to the Boltzmann factor $N_1\exp[-U(k_BT)^{-1}]$, and for the inverse transition $2\rightarrow1$ (v^-) the probability is proportional to $N_2\exp[-U(k_BT)^{-1}]$.

Suppose a shear stress τ causes a linear shift $|\tau|\Omega$ of the energy level of the first state and a shift $-|\tau|\Omega$ of the second state, Ω being the activation volume. Then the differential equation for the occupation of state 1 is given by

$$\frac{dN_1}{dt} = -N_1 v_0 \exp\left[-\frac{(U-|\tau|\Omega)}{k_BT}\right] + (N-N_1) v_0 \exp\left[-\frac{(U+|\tau|\Omega)}{k_BT}\right] \quad (120)$$

where v_0 is the frequency associated with the motions of chain segments at a temperature T. We assume that at $t=0$, $N_1=N_2=N/2$. The relaxation time δ of the transition is given by

$$\frac{1}{\delta} = v_0 \left\{\exp\left[-\frac{(U-|\tau|\Omega)}{k_BT}\right] + \exp\left[-\frac{(U+|\tau|\Omega)}{k_BT}\right]\right\} \quad (121)$$

Because the energy shift $\tau\Omega$ is defined to be a positive quantity, the absolute value of τ is introduced in the exponential function.

The activated transition will now be described in its simplest form, viz. for $U(k_BT)^{-1}\gg1$. This implies that at low temperatures even for a small stress the backward transition rate v^- can be neglected with respect to forward transitions v^+ and the relaxation time can be approximated by

$$\frac{1}{\delta} = v_0 \exp\left[-\frac{(U-|\tau|\Omega)}{k_BT}\right] \quad (122)$$

Hence, Eq. 120 becomes

$$\frac{dN_1}{dt} = -\frac{N_1}{\delta} \quad (123)$$

For constant $|\tau|$ the solution of this equation is given by

$$N_1 = \frac{N}{2} \exp\left(-\frac{t}{\delta}\right) \tag{124}$$

As for the derivation of Eqs. 122, 123 and 124 only the transitions 1→2 have been counted, these equations do not describe recovery processes, where the transitions 2→1 are important as well. These approximations have been made for convenience's sake, but neither imply a limitation for the model, nor are they essential to the results of the calculations. Equation 124 is the well-known formula for the relaxation time of an Eyring process. In Fig. 65 the relaxation time for this plastic shear transition has been plotted versus the stress for two temperature values. It can be observed from this figure that in the limit of low temperatures, the relaxation time changes very abruptly at the shear yield stress $|\tau_y|=U_0/\Omega$. Below this stress the relaxation time is very long, which corresponds with an approximation of elastic behaviour.

The transition from ideal elastic to plastic behaviour is described by the change in relaxation time as shown by the stress relaxation in Fig. 66. The immediate or plastic decrease of the stress after an initial stress σ_0 is described by a relaxation time equal to zero, whereas a pure elastic response corresponds with an infinite relaxation time. The relaxation time becomes suddenly very short as the shear stress increases to a value equal to τ_y. Thus, in an experiment at a constant stress rate, all transitions occur almost immediately at the shear yield stress. This critical behaviour closely resembles the ideal plastic behaviour. This can be expected for a polymer well below the glass transition temperature where the mobility of the chains is low. At a high temperature the transition is a

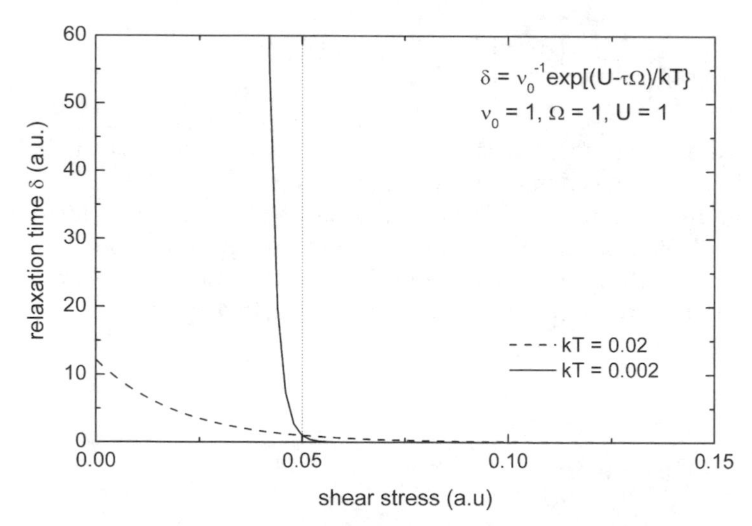

Fig. 65 The relaxation time of an Eyring process as a function of the stress, for a low and a high temperature with $|\tau_y|=U_0/\Omega=0.05$

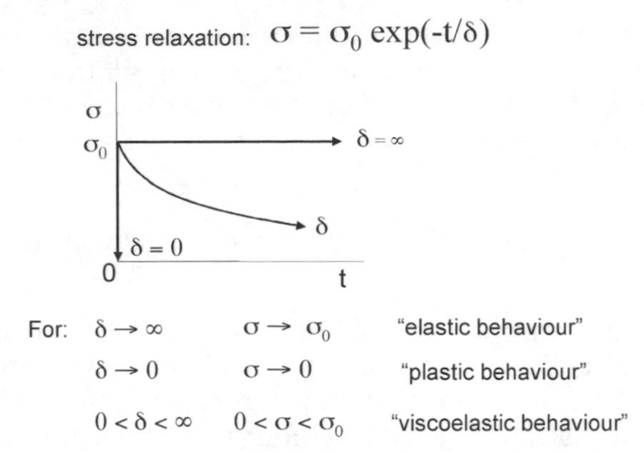

Fig. 66 The elastic, viscoelastic and plastic behaviour can be described by changing the relaxation time δ

smoother function of the applied stress. This behaviour may be expected for a polymer close to the glass transition temperature. In this case the backward transitions should also be taken into account. This different character in the two temperature limits renders the Eyring process very suitable for modelling the coupled yield and viscoelastic deformation of fibres in the glassy state.

If the applied shear stress varies during the experiment, e.g. in a tensile test at a constant strain rate, the relaxation time of the activated transitions changes during the test. This is analogous to the concept of a reduced time, which has been introduced to model the acceleration of the relaxation processes due to the deformation. It is proposed that the reduced time is related to the transition rate of an Eyring process [58]. The differential Eq. 123 for the transition rate is rewritten as

$$\frac{dN_1}{dt_u} = -N_1 \tag{125}$$

with the reduced time t_u given by

$$dt_u = \frac{dt}{\delta} \tag{126}$$

Thus

$$t_u = \nu_0 \int_0^t \exp\left[-\frac{(U - |\tau(t')|\Omega)}{k_B T}\right] dt' \tag{127}$$

It is proposed that the viscoelastic and plastic shear strain of a domain, $\varepsilon_{13}^{v}(t) = \tan(\theta(t) - \theta_0)$, is proportional to $(N/2 - N_1)$. Then it follows from Eqs. 125 and 126 that

$$\varepsilon_{13}^{v}(t) = \tfrac{1}{2} I[1 - \exp(-t_u)] \tag{128}$$

The magnitude of the activated transition is denoted by I, where $I=cN$ and c is an arbitrary constant. A transition density function is introduced to describe the viscoelastic and plastic shear deformation of the domain. Hence, following Eq. 107 the total shear strain of a domain in terms of the ERT model is given by

$$|\varepsilon_{13}(t)| = |\tan\theta(t) - \Theta)| = \frac{|\tau(t)|}{2g} + \tfrac{1}{2}\int_0^\infty I(U)\,[1 - \exp(-t_u)]\,dU \tag{129}$$

and t_u is calculated from the loading history with Eq. 127. The relation between the fibre strength and the time can be obtained from load rate measurements. In order to derive an analytical expression for the strength as a function of the loading rate λ, the integral in Eq. 127 defining the reduced time has to be evaluated. For $\sigma=\lambda t$ the shear stress becomes

$$|\tau(t)| = \lambda t \sin\theta(t)\cos\theta(t) \approx \frac{2g\lambda t\tan\Theta}{2g + \lambda t} \tag{130}$$

As an approximation it is assumed that $\lambda t \ll 2g$ or $|\tau(t)| \approx \lambda t\tan\Theta$. Furthermore the high-strain approximation for the transition density function will be applied, viz. $I(U)=I_0$ on the interval $[U_0, U_m]$ and $I(U)=0$ elsewhere [10]. Equation 129 then yields

$$|\varepsilon_{13}(t)| = \frac{\lambda t\tan\Theta}{2g} + \frac{I_0 k_B T}{2}\left[\chi + \log\left(\frac{\nu_0 k_B T}{\lambda\Omega\tan\Theta}\right) + \frac{(\lambda t\Omega\tan\Theta - U_0)}{k_B T}\right] \tag{131}$$

with the Euler constant $\chi \approx 0.57722\ldots$ and $k_B=1.38\cdot 10^{-23}$ JK^{-1}. From Eq. 131 an expression for the yield stress in the tensile curve of a polymer fibre can be derived. For the shear yield stress of a domain the following condition holds

$$|\tau_y| = 2g\varepsilon_{13}^{y} \tag{132}$$

which implies that the second term on the right-hand side (within the brackets) of Eq. 131 equals zero. The introduction of $\sigma_y=\lambda t$ results then for the yield stress in the tensile curve of a polymer fibre in

$$\sigma_y = \left[-\chi + \log\left(\frac{\lambda\Omega\tan\Theta}{\nu_0 k_B T}\right)\right]\left(\frac{k_B T}{\Omega\tan\Theta}\right) + \frac{U_0}{\Omega\tan\Theta} \tag{133}$$

Equation 133 is similar to the formula for the strain and temperature dependence of the yield point calculated with the thermally activated viscosity proposed by Eyring and Bauwens [37, 59].

A reference temperature T_0 is introduced, which is the equivalent of the "Vogel" temperature and found empirically in the field of polymer viscosity [60]. Thus in Eq. 131, T is replaced by $(T-T_0)$. Assuming that the limiting shear strain for fracture equals β, the fracture condition for the load rate experiment becomes

$$\beta = \frac{\lambda t_b \tan\Theta}{2g} + \tfrac{1}{2} I_0 k_B (T - T_0) \left\{ \log\left[\frac{v_0 k_B (T - T_0)}{\lambda\Omega\tan\Theta}\right] + \frac{(\lambda t_b \tan\Theta - U_0)}{k_B (T - T_0)} + \chi \right\} \tag{134}$$

which results for the strength, $\sigma_b = \lambda t_b$, in the following linear function of $\log\lambda$

$$\sigma_b = C_1 \log\lambda + C_1 \left[\frac{\frac{2\beta}{I_0} + U_0}{k_B (T - T_0)} + \log\left(\frac{\Omega\tan\Theta}{k_B (T - T_0)}\right) + C_2 \right] \tag{135}$$

where

$$C_1 = \frac{g I_0 k_B (T - T_0)}{(1 + g I_0 \Omega)\tan\Theta} \quad \text{and} \quad C_2 = -\log v_0 - \chi$$

As a validation experiment a large number of load speed measurements on Twaron 2200 PpPTA yarn at different temperatures have been carried out. In order to limit the scatter of the data a slight twist was applied to the yarn. Figure 67 shows the fit of the linear relation Eq. 135 with the experimental data. The values for the parameters used in this fit are listed in Table 5. As stated earlier, the linear relationship Eq. 135 was derived for the approximation $\lambda t \ll 2g$. According to Eq. 134, for large values of the load rate the second term should become very small. Indeed, in Fig. 67 the observed data tend to level off for

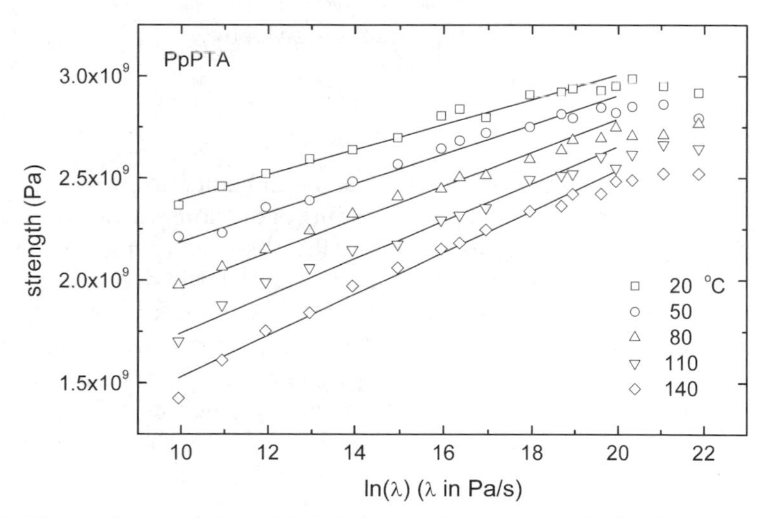

Fig. 67 Observed strength data of PpPTA (Twaron) yarns versus the load rate measured at 20, 50, 80, 110 and 140 °C. The *drawn lines* were calculated with Eq. 135 using the parameters in Table 5

Table 5 The parameters used in the fitting of Eq. (135) to the observed data in Fig. 67

T (K)	β	$\tan\Theta$	g (10^9 Nm^{-2})	I_0 (10^{20} J^{-1})	U_0 (10^{-20} J)	Ω (10^{-30} m^3)	T_0 (K)	$\log v_0$
293	0.08	0.15	1.8	0.65	13.3	250	121	9.4
323	0.08	0.15	1.65	0.65	13.5	255	121	9.4
353	0.08	0.15	1.52	0.65	13.3	255	121	9.0
383	0.08	0.15	1.41	0.65	13.2	250	121	9.4
413	0.08	0.15	1.3	0.65	13.3	255	121	9.4

large values of $\log\lambda$. This observation leads to the conclusion that the fibre can break in either a viscoelastic or in an elastic regime. However, the strength given by Eq. 135 is for shear failure only. As we discussed in Sect. 2.5, for the elastic regime the chain extension has to be taken into account resulting in the molecular composite model.

4.4 Lifetime Relationship Derived from the Eyring Reduced Time Model

The Eyring reduced time model provides the framework for the derivation of the creep equation of polymer fibres [10]. The creep shear strain of a domain is given by

$$|\varepsilon_{13}(t)| = |\tan\theta(t) - \Theta)| = \frac{|\tau(t)|}{2g} + \tfrac{1}{2}\int_0^\infty I(U)\,[1 - \exp(-t_u)]\,dU \qquad (129)$$

The reduced time for the creep experiment is given by

$$t_u = v_0 \int_0^t \exp\left[-\frac{(U - |\tau|\Omega)}{k_B T}\right] dt \qquad (136)$$

It is assumed that the shear stress is constant and equal to $|\tau|=\sigma\sin\theta_0\cos\theta_0$, where θ_0 is the angle immediately after loading of the fibre with a stress σ. This is an approximation, because as shown by the observed change of the sonic modulus during creep, the chain orientation angle slowly decreases. The observed stress dependence of the creep rate of PpPTA fibres for a creep stress up to about 2 GPa is given by Eqs. 108 and 109. For higher stresses no experimental data are available. This observed dependence can also be derived from Eq. 129 by employing the linear transition density function $I(U)=I_1 U$, with I_1 being a constant [10]. As shown in ref. [10] this yields for the creep rate

$$\frac{d\varepsilon_f(t)}{d\log(t)} = \frac{I_1 \Omega k_B T}{2} \frac{\sigma \sin^2\Theta}{\left(1 + \frac{\sigma}{2g}\right)^3} \qquad (137)$$

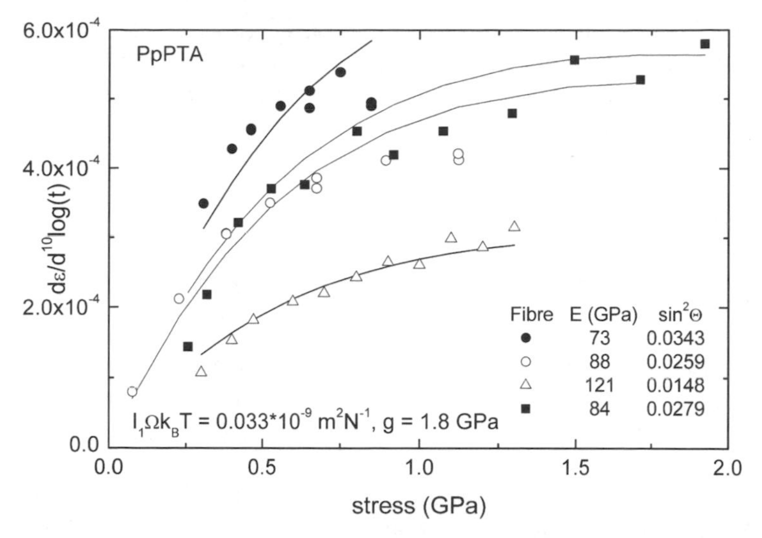

Fig. 68 Comparison of the observed creep rate of PpPTA fibres with the curves calculated with Eq. 137 using $j_1=I_1\Omega k_B T=0.033\cdot 10^{-9}$ m^2N^{-1}[7]

This result can also be derived from Eqs. 10, 108 and 109 with $j_1=I_1\Omega k_B T$. Figure 68 shows a good agreement between the observed and calculated creep data of PpPTA fibres [7].

It can be shown that for σ>2 GPa a constant transition density function $I=I_0$ yields almost the same stress dependence of the creep rate as the linear function. Therefore, in order to keep the calculations tractable we derive the lifetime of a fibre by applying the same density transition function as was used in the calculation of the dependence of the strength on the load rate, viz. $I(U)=I_0$ on the interval $[U_0, U_m]$ and $I(U)=0$ elsewhere. This results for the shear strain of a domain in

$$|\varepsilon_{13}(t)| \approx \frac{|\tau|}{2g} + \tfrac{1}{2} k_B T I_0 \left[\log(\nu_0 t) + \frac{(|\tau|\Omega - U_0)}{k_B T} + \chi \right] \tag{138}$$

with $\chi \approx 0.57722...$ being the Euler constant. The lower boundary U_0 can be considered as the threshold or activation energy of the creep.

Assuming that the fibre breaks due to shear failure, the fracture condition $|\varepsilon_{13}(t_b)|=\beta$ is now applied

$$\beta = \frac{|\tau|}{2g} + \frac{kTI_0}{2} \left[\log(\nu_0 t_b) + \frac{|\tau|\Omega - U_0}{kT} + \chi \right] \tag{139}$$

Equation 139 yields a linear relation between the shear stress $|\tau|=\sigma \sin\theta_0 \cos\theta_0$ and $\log t_b$. However, as can be deduced from Eq. 114, the shear stress is not a linear function of the fibre stress σ. Application of Eq. 114 in Eq. 139 results in a non-linear lifetime equation. However, in the stress range applied in the lifetime experiments, this deviation from linearity is small. Moreover, the change of the angle θ_0 over the stress range applied in the lifetime measurements is relatively small. Therefore, as an approximation for well-oriented fibres the shear stress is taken to be $\tau=-\sigma\sin\theta_a$, where θ_a is the average value of θ_0 in the applied stress range. This results in the following linear equation of the lifetime of a polymer fibre in a creep failure experiment

$$^{10}\log t_b \approx \frac{2\beta + I_0 U_0}{2.3 \cdot I_0 k_B (T - T_0)} - \frac{\chi}{2.3} - {}^{10}\log \nu_0 - \frac{[\Omega + (g I_0)^{-1}] \sin\theta_a}{2.3 \cdot k_B (T - T_0)} \tag{140}$$

Since the linear dependence of the shear stress on the fibre stress has also been applied in the derivation of the load rate Eq. 135, the parameters in this equation can be compared with the parameters obtained from the lifetime relationship.

Over the whole stress range from zero onwards, the lifetime relation should be a non-linear relation between log$t$ and σ because as the stress becomes zero the lifetime should go to infinity. Therefore Eq. 140 only holds within the constraints used in the derivation. Wu has measured the lifetimes of PpPTA filaments at three temperatures, viz. 21, 80 and 130 °C. These data have been fitted with Eq. 140 using the parameters derived from the load rate measurements as a starting value. In Fig. 69 the data of Wu are represented by dashed lines which have been obtained from linear regression of the individual data [30]. In the fitting procedure a slight variation in the values of θ_a, U_0, Ω and ν_0 was necessary to obtain a good fit. The parameters of the drawn lines are listed in Table 6 and

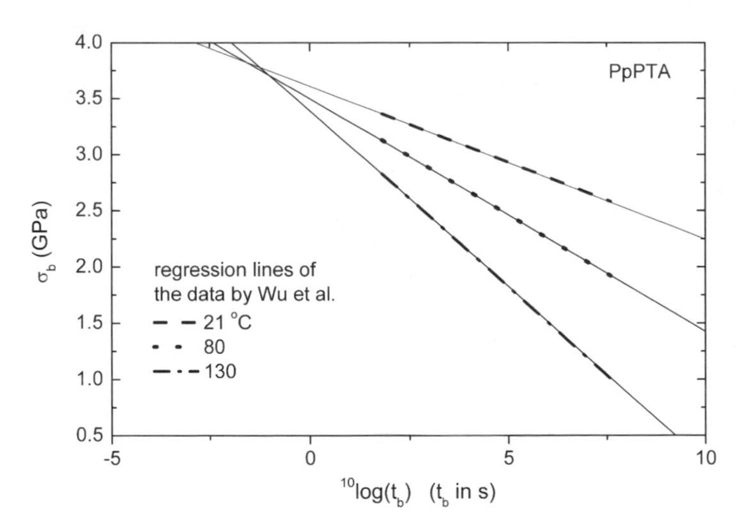

Fig. 69 The lifetime data of Wu et al. are represented by their regression lines and indicated in this figure by *dashed and dotted lines*. The *continuous lines* are calculated with Eq. 140 using the parameter values listed in Table 6

Table 6 The parameter values of the lines defined by Eq. (140) and drawn in Fig. 69

T (K)	β	$\sin\theta_a$	g (10^9 Nm^{-2})	I_0 (10^{20} J^{-1})	U_0 (10^{-20} J)	Ω (10^{-30} m^3)	T_0 (K)	$\log\upsilon_0$
293	0.08	0.142	1.8	0.653	13.3	275	121	4.72
353	0.08	0.142	1.5	0.653	11.3	240	121	3.34
403	0.08	0.135	1.3	0.653	10.3	200	121	1.61

show only small differences with the parameter values in Table 5. Full agreement of the two sets of parameters cannot be expected, because the data by Wu are based on the strength of Kevlar 49 PpPTA filaments and the data of the load rate measurements are from Twaron PpPTA yarns. However, although the filaments and the yarns are from different manufacturers, the agreement between the parameters derived from the load rate measurements and from the lifetime measurements is satisfactory.

According to the simple Eq. 115 and the full Eq. 140, the lifetime of a fibre measured at a constant load decreases with increasing orientation parameter. The dependence of the slope of the curve, $\log(t_b)$ vs σ_b, on the initial orientation distribution has been calculated for PpPTA fibres using Eq. 140. Figure 70 shows that at constant load for increasing orientation angle the lifetime curves become steeper, while at the same time the lifetime decreases. This effect has been observed for nylon 66 yarns as shown in Fig. 71, where the lifetime data

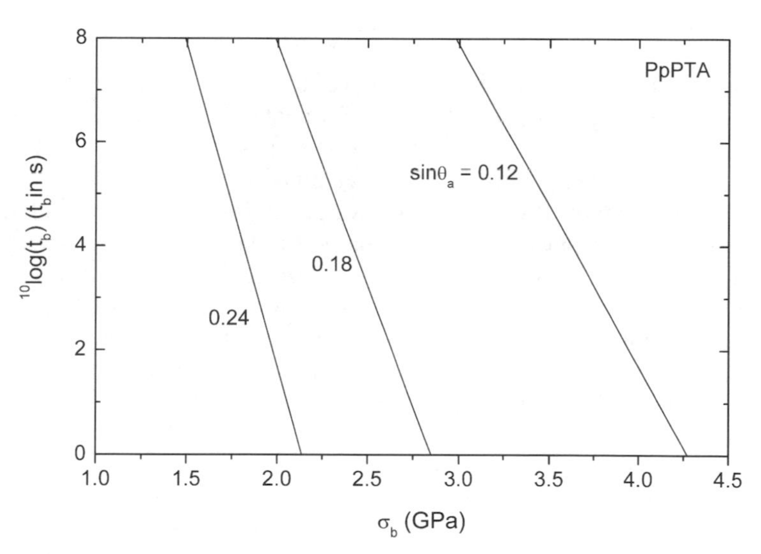

Fig. 70 Calculated lifetime curves of PpPTA fibres for three orientation angles using Eq. 140 with the parameters T=294 K, g=1.8 GPa, U_0=13.3·10^{-20} J, I_0=0.653·10^{20} J^{-1}, T_0=121 K, Ω=275·10^{-30} m^3, $\log\nu_0$=4.72

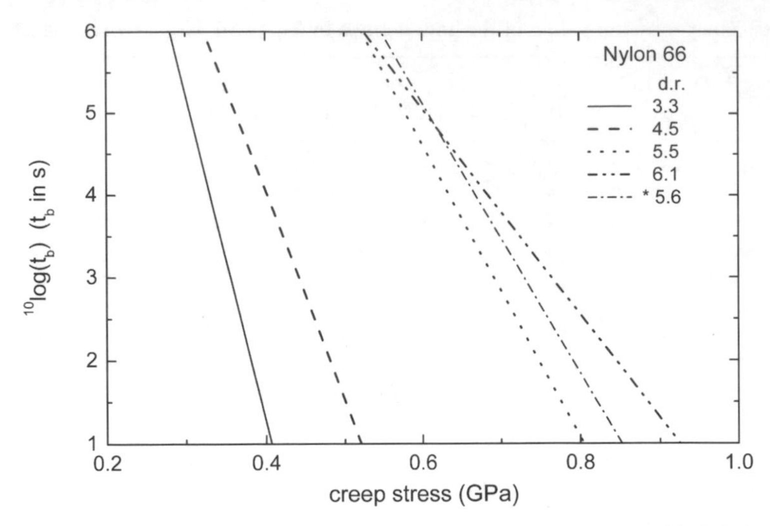

Fig. 71 Lifetime curves as a function of the creep load for nylon 66 yarns with different draw ratios (*d.r.*). The yarn with a stretch ratio of 5.6 is from a different polymer batch [54]. The *drawn lines* represent the regression lines of the observed data

are shown for a series yarns made with different draw ratios [54]. The effect has also been observed for polyacrylonitrile yarns [61].

4.5 Conclusions

The calculated load rate and lifetime curves using the criterion of the critical shear strain are in agreement with the experimental data. However, as discussed in Sect. 2.5, the molecular composite model shows that the fibre strength obtained in a normal tensile test is also determined by the longitudinal strength σ_L due to the action of the normal stress $\sigma\cos^2\theta$. Neglect of this contribution results in strength values that are calculated too large. Thus the strength calculated with the ERT model for large load rates and short lifetimes is larger than the observed values. Indeed, this can be observed in Fig. 67 where the experimental data level off for high load rates. Further confirmation is found in the observations made by Wu et al. [30]. These authors presented results of lifetime and tensile tests of PpPTA filaments measured at 21, 80 and 130 °C. The strength values obtained from the tensile tests were considerably lower than the values derived from the extrapolation of the observed lifetime relation to very brief time periods. Moreover, the filaments that failed in creep after very short time periods (<3 min) seemed to have failed through transverse crack propagation; over one or two fibre diameters along the fibre no long-range splitting or fibrillation was observed. Apparently this indicates a change in failure mode along the time axis in a creep failure experiment. The authors concluded that the short-term strength as determined from a normal tensile test is not a good predictor of long-term life, which is confirmed by the proposed model for creep failure.

The presented derivations of the load rate and the lifetime relationships applying the shear failure criterion are based on a single orientation angle Θ for the characterisation of the orientation distribution. Therefore these relations give only an approximation of the lifetime of polymer fibres. Yet, they demonstrate quite accurately the effect of the intrinsic structural parameters on the time and the temperature dependence of the fibre strength.

The strength of a fibre is not only a function of the test length, but also of the testing time and the temperature. It is shown that the introduction of a fracture criterion, which states that the total shear deformation in a creep experiment is bounded to a maximum value, explains the well-known Coleman relation as well as the relation between creep fracture stress and creep fracture strain. Moreover, it explains why highly oriented fibres have a longer lifetime than less oriented fibres of the same polymer, assuming that all other parameters stay the same.

Finally, a new and interesting application of the ERT model can be the modelling of fatigue. In fatigue testing a varying load is applied on the fibre. The dynamic nature of the loading pattern with angular frequency ω is described by introducing a sinusoidal stress function $\sigma=\sigma_0\sin\omega t$ in the shear stress $|\tau|=\sigma\sin\theta\cos\theta$ of the basic Eq. 120. In addition the backward transitions characterised by ν^- should be taken into account.

5 Concluding Remarks

5.1 Discussion

It has been shown that, by neglecting the chain extension, a simple theory based on the continuous chain model and the employment of a critical shear stress or shear strain as the fracture criterion provides a good approximation of the shape of the fracture envelope of polymer fibres $\{\varepsilon_b,\sigma_b\}$. However, for highly oriented fibres ($\Theta<\arctan\beta$) it yields an infinite value of the strength. The incorporation of the molecular composite model takes account of the effect of the chain extension on the fibre strength. It gives the ultimate strength, σ_L, of fibres with chains oriented parallel to the fibre axis in terms of the strength of the intermolecular bonds between chains of finite length. Therefore its value is considerably smaller than the breaking strength of the single polymer chain. However, the examples given in Sect. 2.5.2 show that the theoretical estimates of the shear modulus g, and of the ultimate strength derived from the activation energy of creep and from the molecular force field calculations, leave ample room for improvement.

The proposed theory of the tensile strength of polymer fibres provides the relationship between the strength and the modulus, which is in good agreement with the observed relations on PET, POK, cellulose II and PpPTA fibres. Moreover, it also explains the excellent tensile properties of the new rigid-rod polymer fibres PBO (Zylon) and PIPD (or M5). As shown by the examples of PET and POK fibres, the proposed theory can also be applied to the flexible-chain polymers. Therefore, the strong UHMW PE fibres with only weak Van der Waals bonding between the chains require some discussion here. Strong PE fibres can only be made by employing extremely long chains and by creating in the gel spinning process an almost perfect parallel orientation of the chains. For the calculation of the ultimate strength the chain modulus is taken to be e_c=280 GPa and, due to the weak interchain bonds, a comparatively low value of the internal shear modulus g=0.8 GPa is used. An ultimate strength of 6.5 GPa can only be attained by using a critical shear strain β=0.19. This large value of the critical shear strain can be associated with the extremely long chains and the extremely narrow orientation distribution in these fibres. The highest observed modulus and strength of this fibre are E=264 and σ_b=7.2 GPa [62]. Figure 72 shows the strength as a function of the modulus calculated with Eq. 58. The deviating structure and morphology of the UHMW PE fibres compared to the structure of the rigid-rod polymeric fibres may yield experimental relations between the orientation parameter, the modulus and the strength being different from the relationships computed with the molecular composite model [62–64]. However, lack of sufficient experimental data on the orientation of UHMW PE fibres prevents a comparison.

Although the cross-sectional areas of the chains in PpPTA and PE are slightly different, 0.204 and 0.183 nm^2, respectively, there is a considerable dif-

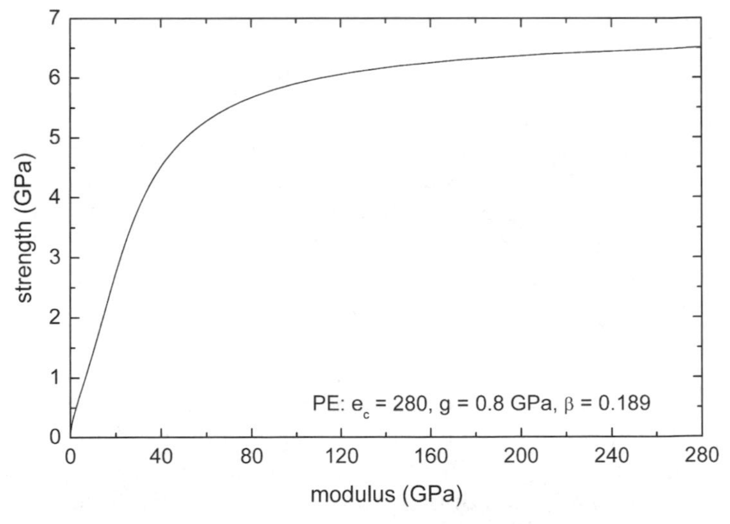

Fig. 72 UHMW PE: the strength as a function of the modulus calculated with Eq. 58 using e_c=280, g=0.8 GPa and β=0.19

ference between the observed strength of these fibres. Part of this difference is probably due to the much higher degree of chain orientation in the UHMW PE fibres as judged from the X-ray diffraction patterns. As has already been mentioned, a considerably larger value of the critical shear strain due to the extremely long chains in UHMW PE may also contribute to the larger strength of these fibres. Moreover, the effect of impurities on the strength may be very different in these fibres. Firstly, due to the flexibility of the PE chain, the stress concentrations due to impurities relax much more rapidly than in PpPTA fibres. Secondly, the process conditions of the PpPTA fibre involving the preparation of a spinning solution in an inorganic acid make it much more likely that the level of impurities is considerably higher in PpPTA fibres than in PE fibres. As shown in Sects. 2.2 and 2.3, the detrimental effect of impurities increases with a broader orientation distribution. Thus, not only the higher level of impurities in PpPTA, but also the wider orientation distribution and, consequently, a smaller Griffith length may be responsible for the lower strength of PpPTA fibre compared to the UHMW PE fibre.

In aramid fibres, such as PpPTA, with a chain modulus of 240 GPa and a mixture of van der Waals and hydrogen bonding, the relatively short chains with a narrow orientation distribution yield an observed modulus of 120 GPa and a strength of 4.5 GPa. For the PBO fibre the large chain modulus of about 500 GPa and the relatively weak interchain bonding yield a large fibre modulus of 280 GPa and a yarn strength of 6 GPa only if the orientation distribution is very narrow. A large chain modulus together with a two-dimensional hydrogen-bonded network, as found in the newly developed PIPD fibre, has resulted so far in a modulus well above 300 GPa and a filament strength of nearly 7 GPa for a contracted orientation distribution.

Of particular significance to the strength of polymer fibres is the outcome of the analysis of the distribution of the strain energy. It demonstrates that due to the chain orientation distribution, the strain energy delivered in a tensile test is preferably channelled into the domains with angles in the tail of the distribution. This implies that even without structural inhomogeneities like impurities, the critical shear strain, β, or the critical shear stress, τ_b, is first exceeded in the domains with angles that are in the tail of the orientation distribution. Consequently, due to the presence of the chain orientation distribution alone, even polymer fibres without any impurities and voids attain a lower strength than the ultimate strength calculated for a fibre with all chains parallel to the fibre axis. Moreover, fibres made of the same polymer having the same values of $\langle \sin^2\Theta \rangle_E$, g and e_c, and thus with equal moduli, may still have a different strength due to a difference in the shape of the orientation distribution.

In principle the proposed model for the strength of polymer fibres can be improved by taking into account the change in shape of the orientation distribution during the extension of the fibre. However, from the fact that the application of only a single orientation angle for the representation of the orientation distribution has yielded the observed relations quite accurately, it follows that such a refinement is likely to give only a marginal improvement. The observed relationships between the fibre strength and the concentration of the spinning solution of PpPTA and DABT have been explained with the proposed strength model in conjunction with Picken's theory describing the order parameter in liquid crystalline solutions of polymers. Table 7 presents a survey of the elastic constants and the estimated maximum tensile strength, σ_L, of various polymer fibres. By using a constant value for β, this list does not take into account that the fracture shear strain may be different for the different kinds of interchain bonding. Yet, it shows that the PIPD (M5) fibre has the potential to achieve the highest tenacity of all organic polymer fibres. Other proven properties of the M5 fibre are the exceptionally good impact strength, the high axial compressive strength and the excellent fire protection properties [8, 65–69].

The PBO fibre demonstrates the importance of the degree of chain orientation. Because there are no hydrogen bonds, the interchain bonding in the PBO is less strong than in the PpPTA fibre. Yet, the observed strength of the PBO fibre is considerably larger than that of PpPTA. As is shown by Figs. 26 and 27, this is mainly due to a combination of the large chain modulus and the narrow orientation distribution of the chains. For PBO the chain modulus is $e_c \approx 500$ GPa and for the as-spun fibre $\langle \sin^2\Theta \rangle_E \approx 0.018$ and for PBO-HT $\langle \sin^2\Theta \rangle_E \approx 0.006$, whereas for PpPTA $e_c = 240$ GPa and for a fibre with a modulus of about 90 GPa $\langle \sin^2\Theta \rangle_E \approx 0.028$.

Kitagawa et al. studied the deformation process in PBO. In order to determine the elastic constants, e_c and g, according to Eq. 10, they plotted the fibre compliances versus the orientation parameter $\langle \sin^2\Theta \rangle$, instead of the strain orientation parameter $\langle \sin^2\Theta \rangle_E$, and found $e_c = 370$ and $g = 4$ GPa [70, 71]. However, their data show a very large spread, implying that linear regression of the data should have resulted in large esd values for the derived values of e_c and g. Moreover, by applying a non-aqueous coagulation process, they were sucessful in making a PBO fibre with a fibre modulus of 360 GPa, which is very close to their

Table 7 Estimates of the ultimate tensile strength for perfectly parallel orientation of the chains in the polymer fibre calculated with the equation $\sigma_L = 2.3\,\beta\sqrt{(g e_c)}$ and $\beta = 0.1$

Fibre	e_c (GPa)	g^a (GPa)	σ_L (GPa)	Highest observed strength
PET	125	1.5	3.1	2.3
Cellulose II conditioned	88	3	3.7	1.8
PpPTA	240	2	5	4.5
PBO	500	2	7.3	7.3
PIPD (M5)[b]	550	6	13	6.6

[a] The determination of g is described in ref. [40].
[b] The PIPD fibre is still in an early stage of development.

estimated value of e_c. According to Eq. 10 this large fibre modulus strongly suggests that the chain modulus of PBO should have a considerably larger value than 370 GPa.

The difference between the observed moduli of PBO and PIPD-HT fibres is caused by the difference in chain orientation and by the difference of the value of the shear modulus g. As the PIPD fibre is in the early stages of development, it can be expected that after complete development the strength will be well above the present highest value of 6.6 GPa. However, the large g value of the PIPD fibre may render it more susceptible to the detrimental effect of inhomogeneities and impurities, because it will probably lead to somewhat higher stress concentrations than those in fibres with weaker interchain forces. However, Eq. 39 shows that a large g value yields a large Griffith length, which is beneficial to the fibre strength.

The study of the effect of the chain length distribution on the fibre strength has demonstrated that the aspect ratio of the elemental building unit is an important parameter. This leads to the question: what is the elemental unit in a polymer fibre? Assuming that in PpPTA fibres the basic element is the chain itself, the model predicts that at the present level of DP ($=z_n$) of 50 the strength increase due to an increase in DP will be very small indeed. As shown in Sect. 3, this also holds for the effect of the elimination of the low molecular weight fraction. If microfibrils are the building blocks, then the mechanical parameters e_c, g, τ_b and the aspect ratio L/D of the model refer to fibrils themselves. Thus, e_c corresponds to the modulus of the fibril, g becomes the modulus for shear between adjacent fibrils, and τ_b the corresponding shear strength. As very little is known about these quantities, this point will not be discussed any further.

The presented model for the calculation of the influence of the molecular weight on the fibre strength does not deal with the possible dependence of the critical shear strength on the molecular weight distribution. Similar to the macrocomposite model, it uses a constant shear strength. For cellulose and aramid fibres it has been shown that g increases with an improved parallel orientation of the polymer chains, corresponding with a more perfect chain packing [2]. However, nothing is known about its effect on the critical shear strain or the shear strength in fibres. From the observation that one can hammer a lot of nails into a piece of wood before its strength in a direction parallel to the wood grains starts to decline, it can be argued that small structural defects have probably little effect on the strength. Yet, in order to gain a better insight into the influence of the various structural parameters on the lifetime relation of polymer fibres, more experiments are required.

The model of Frenkel for the calculation of the shear modulus and the critical shear strength is not particularly suited to polymers, because it does not take into account the chain length distribution. As we have shown, the critical shear strength is the fracture criterion pertaining to tensile tests applying relatively high strain rates, whereas a critical or maximum shear strain is the proper criterion pertaining to the lifetime measurements. It seems plausible to assume that the critical shear strain may also depend on the average chain

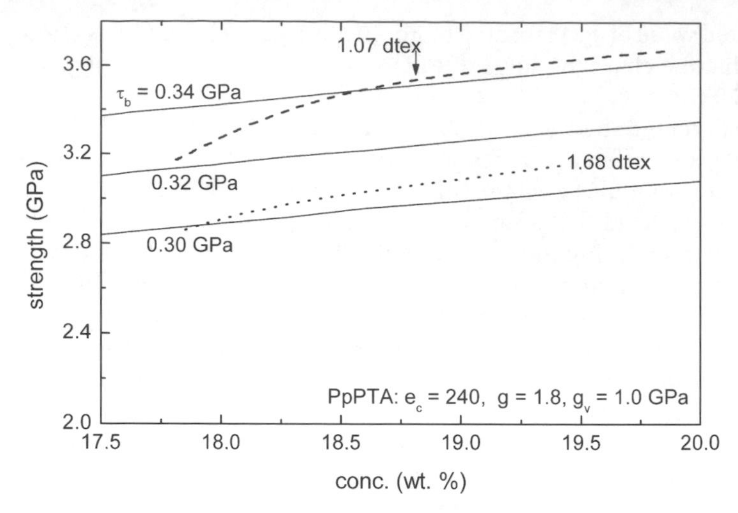

Fig. 73 Strength as a function of the concentration. The *drawn lines* were computed with Eqs. 23, 26, 42, 43, 45, 59, 61a, 61b and 62 for different values of σ_L, and τ_b. The *dotted and dashed lines* are average curves representing the experimental data of normal and microfilament PpPTA yarns, respectively

length in the polymer. Thus, with longer chains the overlap length will increase, presumably raising the critical or maximum shear strain and shear stress. This may be the case in the UHMW PE fibres.

The Eyring reduced time model provides a good description of the creep of PpPTA fibres [10]. Moreover, the application of this model, in conjunction with the maximum shear strain criterion for the fracture of polymer fibres, has been shown to be successful for the interpretation of the observed lifetime relations of the PpPTA fibres and their dependence on the degree of chain orientation and on temperature. The expression of the fibre strength as a function of the load rate given by Eq. 135 demonstrates the importance of the various intrinsic polymer properties for the fibre strength, i.e. a large value of g, β, the threshhold energy for yield U_0 and a small orientation angle Θ will give a higher strength. As shown by the creep failure experiments of Wu et al. on PpPTA, polymer fibres like many other materials can break in either a ductile or brittle manner [30, 72]. This transition of the failure mode is explained by the proposed theory. For example, in Eq. 134 it follows that the ductile or viscoelastic part of the critical shear strain $\varepsilon_{13}^{v}(t) \rightarrow 0$ for $t_b \rightarrow 0$.

Table 8 presents a survey of the basic elastic constants of a series of polymer fibres and the relation with the various kinds of interchain bonds. As shown by this table, the interchain forces not only determine the elastic shear modulus g, but also the creep rate of the fibre.

The importance of the uniformity of structure and morphology for the strength of the fibre is illustrated in Fig. 73. It shows that the observed filament strength of the PpPTA microfilaments is considerably higher than the strength

Table 8 The basic elastic constants g and e_c, the highest filament values of the modulus (E) and the strength (σ_b), together with the average values of the creep compliance ($j(t)$) at 20 °C (ratio of creep rate and load stress) and the interchain bond for a variety of organic polymer fibres

Fibre	e_c (GPa)	g (GPa)	E (GPa)	σ_b (GPa)	$j(t)$ (% dec^{-1} GPa^{-1})	Interchain bond
UHMW PE	280–300	0.7–0.9	264	7.2	0.3	Van der Waals
PET	125	1.0–1.3	38	2.3	0.13	Van der Waals/dipole
Cellulose II	88	2.0–5.0 [a]	55	1.8	0.4 [b]	Bidirectional hydrogen
Cellulose I (native fibre)	136	1.5	110	–	–	Unidirectional hydrogen/van der Waals
PpPTA	240	1.5–2.7	140	4.5	0.02–0.04	Unidirectional hydrogen/van der Waals
PBO	500	1.0–1.5	280	7.3	0.005	Van der Waals/dipole
PIPD [c]	550	6.0	330	6.6	0.0035	Bidirectional hydrogen

[a] Depending on water content.
[b] Conditioned at 65% R.H.
[c] PIPD is a fibre in an early stage of development.

of filaments with a standard count. Thus, an increased uniformity of the structure in the cross section of the fibre, together with a decrease in impurity content, should result in a higher value of τ_b.

A Weibull analysis of the filament properties of a yarn provides the proper information concerning the possible increase in yarn strength by altering extrinsic factors. During spinning of a PpPTA yarn the individual draw ratios of the filaments are not precisely equal, which implies that the filaments of a yarn show a distribution of the modulus. For a yarn without large structural inhomogeneities and impurities the presented theory indicates that the filament strength should be an increasing function of the filament modulus. When due to a large spread the experimental filament data taken randomly from the cross section of a single bundle fail to reveal this relation, the detrimental effect of extrinsic factors such as impurities still dominates the filament strength. However, when for an arbitrary selection of filaments taken from a yarn the theoretical relation between strength and modulus becomes apparent, a further considerable increase in yarn strength at the same modulus cannot be expected.

With regard to the observed variation of the fracture morphology of polymer fibres, the presented analysis of the strength may provide some understanding of its causes, in particular of those fibres made from rigid-rod polymers. Yoon's model assumes that a polymer fibre with the chains perfectly parallel-oriented to the fibre axis fails under tension when the interfacial shear stress exceeds a critical value. This implies shear fracture, though the shear stress, $|\tau|=\sigma\sin\theta\cos\theta$, acting on the domain equals zero. The fracture of a macrocomposite made of short and strong reinforcing fibres, under a tensile stress directed along the fibres, causes a crack surface perpendicular to the fibre direction with individual fibres pulled out from the matrix [35, 73]. In analogy, the tensile fracture morphology of a rigid-rod polymer fibre with a highly contracted chain orientation distribution ($|\tau|=\sigma\sin\theta\cos\theta\rightarrow 0$) is likely to have a more or less brush- or bristle-like appearance. The short brush "hairs" consist of bundles of chains pulled out from the crystallites. Moreover, also in a domain making a finite angle with the fibre axis, Yoon's model implies that the normal stress $\sigma\cos^2\theta$ causes an interfacial shear stress between the chains. Yet, for an increasing orientation angle the effect of the shear stress on the domain becomes more important. Therefore, the fracture morphology of rigid-rod polymer fibres is expected to change along the fracture envelope of a polymer. Broken low- and medium-oriented fibres will show split fibres and a more or less fibrillar fracture surface, whereas broken highly-oriented fibres tend to show a more or less brush-like appearance associated with brittle fracture. A similar change of the fracture morphology is likely to occur in a creep failure experiment on a well-oriented fibre. Low creep loads corresponding with a long lifetime result in a fibrillar fracture morphology, whereas high creep loads associated with very short lifetimes will show elastic and brittle failure with a brush- or bristle-like fracture surface.

Also, other factors may influence the fracture morphology as is demonstrated by the following examples. Glass fibres having an isotropic structure and mechanical properties display brittle fracture, whereas anisotropic wood fibres show a fibrillar fracture morphology. Hence, the anisotropy of the dimensions and the elastic constants of the elemental building block are likely to have some influence. In rigid-rod polymer fibres, which are made in a wet-spinning process, the micro-morphology consists of long fibrils separated by very fine elongated voids, which facilitate crack growth by shear failure. Moreover, the mechanical anisotropy of the domain in these fibres as expressed by the ratio e_c/g is considerable: for cellulose II 20, PIPD 90, PpPTA 120 and PBO 200. This range may explain the observation that on the one hand cellulose II fibres show a fibrillated fracture surface with short fibrils or a brush-like appearance, whereas on the other hand medium-oriented PpPTA and PBO fibres show split fibres and long fibrillated fracture surfaces. The anisotropy expressed by e_c/g in polymer fibres is caused by the strong covalent bonds within the chain and the weak viscoelastic secondary bonds between the chains. Presumably this anisotropy in bonding strength also causes the difference between the fracture morphology of PAN- and pitch-based fibres. Although both fibres break in a brittle mode, close inspection reveals subtle differences in the fracture surface.

Non-graphitised PAN-based carbon fibres consist of irregularly bent graphite-like planes, which are oriented to some degree along the fibre axis and are laterally cross-linked by strong sp^3–sp^3 bonds resulting in a modulus of shear between the graphite planes of about 30 GPa and $e_c/g \approx 35$ [74]. Except for marked features caused by holes or particles, tensile fracture of these fibres shows a more or less smooth fracture surface transverse to the fibre axis. This is in agreement with the conclusion made by many investigators that PAN-based fibres have a non-fibrillar structure. Pitch-based carbon fibres consist of highly ordered layers of graphite planes well-oriented parallel to the fibre axis and often showing a turbostratic structure. Some meso-phase pitch-based fibres possess an additional degree of order corresponding with the development of the aromatic layers into a stacking order with the crystal structure of graphite. The planes are laterally bonded by very few covalent bonds yielding a shear modulus of about 5 to 10 GPa and $e_c/g \approx 120$. The fracture morphology of these fibres is characterised by a rough surface normal to the fibre axis, which has been described as of "pulled-out" appearance [75–77]. The roughness is due to a sheet-like morphology originating from sheets consisting of stacks of graphite planes.

5.2 Can Post-drawing Improve the Fibre Strength?

High-modulus PpPTA fibres are made by post-drawing a medium-modulus fibre at elevated temperatures. In order to understand the effect of this process step, the mechanism of drawing polymer fibres is discussed. Drawing of a polymer fibre below its glass transition temperature T_g is called "cold drawing". At T_g, the interchain bonding becomes severely weakened by the onset of large

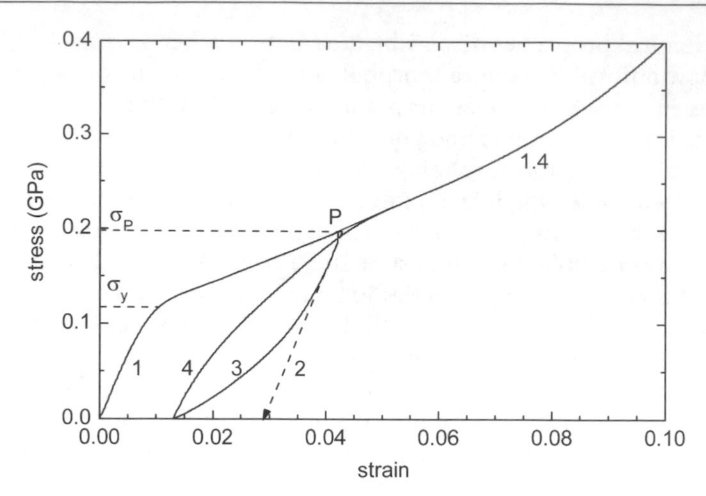

Fig. 74 Schematic representation of the tensile curve of a fibre during (1) first loading, (2) hypothetical elastic unloading, (3) unloading and (4)second loading. Note that the second part of curve 4 practically coincides with curve 1 [1, 6]

segmental motions of the polymer chain, and a transition from neck (yielding) to uniform deformation occurs [78]. In general, drawing above T_g creates a fibre with a structure and morphology that is very different from that of the original fibre. This is due to the combined action of reorientation and recrystallisation processes. Figure 74 depicts a schematic representation of the tensile curve of a polymer fibre measured below T_g [1]. The following curves can be distinguished:

a) Curve 1, coinciding from point P onwards up to fracture with the second part of curve 4, represents the first loading curve; it is characterised by yielding at a strain between 0.5 and 2.5% depending on the orientation of the fibre.
b) Curve 3 is observed during unloading of the fibre from a stress σ_P at point P, well above the yield stress σ_y;.
c) The first part of curve 4 up to P represents the second loading curve after unloading of the fibre along curve 3; it is followed by the second part of curve 4, which coincides with the second part of curve 1. A considerable change of the slope of curve 4 at P is noticed.

When the unloading time at the bottom end of curve 3 is short, the initial slope of curve 4 is slightly steeper than the initial slope of curve 1. This indicates that, due to the first loading to a stress σ_P well above the yield stress, a contraction of the orientation distribution has taken place, resulting in an increase of the initial modulus. The unloading curve 3, together with the first part of curve 4 up to σ_P, is the hysteresis curve followed during cyclic loading between 0 and σ_P. This schematic representation of the tensile behaviour holds for all polymer fibres at temperatures below the glass transition temperature. In other words, it holds for all fibres showing yielding and it can be explained by the sequential orientation mechanism [1, 6]. Depending on the kind of fibre, the reappearance of the yield in the first stage of curve 4 is observed after a long unloading time. As soon as the stress during the second loading is above the maximum stress reached in the previous first loading cycle, the tensile curve of the second loading will follow the first loading curve 1 of the fibre up to fracture.

A general observation is that cold drawing of polymer fibres results in a higher modulus, but not in a higher strength. The sequential orientation mechanism provides the explanation of this phenomenon. The fibre is considered as a parallel arrangement of identical fibrils. Each fibril is a series arrangement of domains composed of parallel packed chains. Hence the tensile extension of the fibre is described by the tensile behaviour of the fibril. The chain axes of the domains in a fibril follow an orientation distribution, as is shown in Fig. 75. The fibre elongation is brought about by elastic chain stretching and by chain rotation as a result of the shear deformation. This chain rotation is caused by the resolved shear stress $|\tau|=\sigma\sin\theta\cos\theta$, which varies from domain to domain in the fibril. For small strains the fibre response is practically elastic, but at the yield strain the chain rotation has a plastic contribution, presumably caused by the temporary loosening of the interchain secondary bonds. For simplicity it is assumed that a pure plastic rotation of the chain occurs when the shear stress is equal to or larger than the shear yield stress τ_y. The value of the shear yield stress is given by $|\tau_y|=fg$, where $0.04<f<0.05$ [1]. In a fibril the maximum shear stress will act on domains with chains having the largest orientation angle θ. Hence, during first loading of the fibre the chains with angles in the range

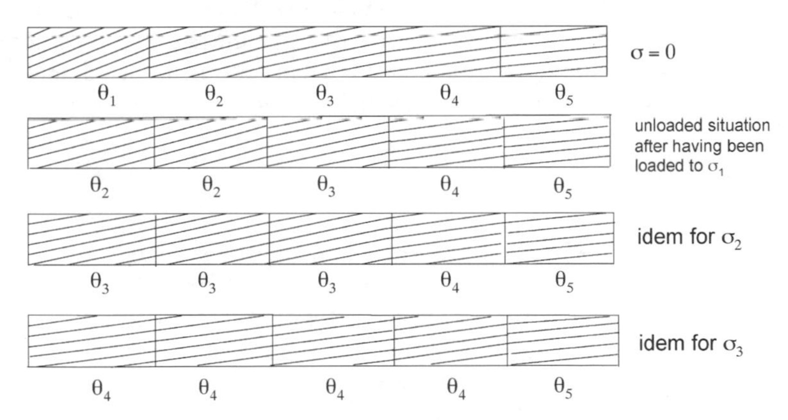

Fig. 75 Schematic representation of the yielding process in a fibril due to sequential plastic orientation of the domains with increasing stress σ. For the domain in the fibril with an angle θ_i, yielding occurs when $\sigma\sin\theta_i\cos\theta_i\geq|\tau_y|$. After loading to a stress $\sigma_1>\sigma_y$ the chains in the domain with the initial angle θ_1 are plastically rotated to the angle θ_2; after loading to a higher stress σ_2 the chains in the domains with the initial angles θ_1 and θ_2 are plastically rotated to the angle θ_3 etc.

$\theta_y<\theta<\pi/2$ will undergo a plastic rotation down to an angle θ_y at the yield stress σ_y. This angle is given by the yielding condition

$$\sigma_y \sin\theta_y \cos\theta_y = |\tau_y| \tag{141}$$

or

$$\theta_y = \tfrac{1}{2}\arcsin(2|\tau_y|/\sigma_y) \tag{142}$$

When the stress is increased to a value of $\sigma_P>\sigma_y$ the chain axes with angles $\theta_P<\theta<\theta_y$ are plastically rotated down to the angle θ_P given by the equation

$$\sigma_P \sin\theta_P \cos\theta_P = |\tau_y| \tag{143}$$

The serial arrangement of domains implies a sequential plastic orientation of the domains in the fibril, i.e. plastic rotation starts at the most disoriented domain and with increasing stress the domains with smaller angles follow. Let us say that in Fig. 75 $\sigma_1=\sigma_P$ and $\theta_2=\theta_P$, then during the first loading curve all domains with angles larger than θ_P have been subjected to elastic and plastic shear deformation. If after complete unloading from the stress σ_P the second loading is started, no plastic rotation can occur for a stress $\sigma<\sigma_P$ because the domains with angles $\theta\geq\theta_P$ allowing a shear stress $|\tau|\geq|\tau_y|$ are absent. Owing to the permanent contraction of the orientation distribution caused by the first loading and by the absence of plastic rotation, the upward branch (4) of the second loading curve is steeper than the corresponding first loading section (1). However, when the stress during the second loading attains a value $\sigma>\sigma_P$, the domains with an orientation angle $\theta<\theta_P$ will be subjected to the shear yield stress. Accordingly, the chain rotation will again contain a plastic contribution and as a result the tensile curve will resume the course followed during the first loading, which is practically that of the section beyond P of curve 4 and coincides with the first loading curve. As we have discussed before, fibre fracture occurs when the shear stress attains the critical shear stress value τ_b. This happens for a tensile stress σ_b in a domain with the orientation angle θ_b of the contracted orientation distribution given by

$$\sigma_b = \frac{|\tau_b|}{\sin\theta_b \cos\theta_b} \tag{144}$$

Irrespective of the pre-loading procedure, during the second loading of the fibre there is also always a domain with an angle θ_b for which Eq. 144 holds. Consequently fibre failure will occur at a stress not larger than σ_b, being the strength observed at first loading of the fibre. Thus, even after first loading of the fibre up to 95% of its original strength, the strength reached during the second loading will neither be higher nor smaller. This mechanism of sequential orientation of domains in the fibril consisting of viscoelastic and plastic rotation of the chain axes holds for all polymer fibres showing yielding behaviour.

In the case of PpPTA fibres there is no indisputable evidence for a glass transition. Even at elevated temperatures the fibres display some yielding behaviour, which implies that drawing at these temperatures occurs apparently below the glass transition temperature. Hence, it seems unlikely that drawing at these temperatures will change the orientation distribution in such a way that the largest orientation angle of the contracted distribution is considerably smaller than the angle θ_b determining fracture during first loading of the fibre. At most, hot drawing may slightly increase the critical shear stress τ_b by bringing about a higher orientation and crystallinity in the domains. This leads us to believe that post-drawing of PpPTA at elevated temperatures does not raise the fibre strength considerably.

There are other effects that will prevent an increase of the fibre strength by a post-drawing process step. Firstly, with increasing temperatures it is observed that due to degradation the strength of polymer fibres decreases rapidly. Secondly, at constant creep stress the lifetime of a fibre decreases with increasing creep temperature, which follows from Eqs. 135 and 140. As the increased thermal motion of the chain reduces the strength of the interchain bonding, a smaller value of g in these equations also reduces the fibre strength for a given lifetime. Thus, it is important to know, for every post-drawing step at elevated temperatures, the lifetime of the fibre at that particular drawing temperature. If, for example, the lifetime of a PpPTA fibre for a stress of 0.3 GPa at 200 °C is 2 s, then a residence time of the yarn in the process of about 0.5 s during a post-drawing step at this temperature will have a negative effect on the strength. Note that this decrease is solely due to accelerated creep described by the Eyring reduced time model. Furthermore, the lifetime relation refers to strength values that have been measured at a particular test length. Therefore, one should be aware that the apparent or corresponding "test" length in the drawing step of the process can be very different from the test length used in the experiment that yielded the lifetime relation. In conclusion, it seems plausible to assume that during the post-drawing stage at elevated temperatures a creep rupture process is in progress. This effect and the chemical degradation are additional causes for the absence of a substantial strength increase in a post-drawing stage at elevated temperatures. So, the residence time of the fibre in any post-drawing process step applied in a polymer fibre production process should be considerably smaller than the lifetime of the fibre at the temperature of the post-drawing step.

The tensile curve of a polymer fibre is characterised by the yield strain and by the strain at fracture. Both correspond with particular values of the domain shear strain, viz. the shear yield strain $\varepsilon_{13}^{y}=f/2$ with $0.04<f<0.05$ or a rotation angle of $\Theta-\theta_y=f/2$ and the critical shear strain $\Theta-\theta_b=\beta$ with $\beta\approx0.1$. For a more fundamental understanding of the tensile deformation of polymer fibres it will be highly interesting to learn more about the molecular phenomena associated with these shear strain values.

5.3 Effect of Residual Stress, Chemical Impurities and Degradation

In the preceding section it has been shown that it is unlikely that a post-drawing step improves the fibre strength. However, in the case of rigid-rod polymer fibres the notable absence of a clear glass transition temperature does not rule out the possibility that post-drawing of these fibres at elevated temperatures cannot be considered as a cold drawing process. The lack of strength improvement in a post-drawing step may also be due to residual stresses.

The presence of impurities and inhomogeneities, as well as a difference in degree of chain orientation between skin and core of a filament, may give rise to residual stresses. Thus, loading and subsequent unloading of a fibre at temperatures below the glass transition temperature can induce residual stresses in the same way as in the case of metals. Presumably, in rigid-rod polymer fibres the mobility of the chains at these temperatures is too small for complete relaxation of local residual stresses caused by loading of the fibre. Although internal or residual stresses have attracted great interest, no definite proof has been given that residual stresses are found in polymers, in particular in polymer fibres. Perhaps, laser Raman spectroscopy offers a tool for measuring small residual stresses. As already indicated in Sect. 2.1, a residual stress implies a residual shear stress, which has the same effect as a reduction of the critical shear stress. When the sum of the residual shear stress and the applied shear stress exceeds the critical shear stress τ_b, the fibre will break. The stress relaxation rate at room temperature of a PpPTA fibre at a stress of 0.65 GPa is about 0.04 GPa per decade. Assuming a constant relaxation rate, this implies complete relaxation of a residual stress of 0.6 GPa after 15 decades of time. Due to this very low relaxation rate, it is plausible that residual stresses brought about by post-drawing do occur.

However, at first sight there seems to be a strong argument against the presence of residual stresses. A first rapid loading of a fibre up to about 95% of the expected strength does not result in a lower strength observed in a second tensile loading immediately after the first loading. In view of the creep rupture process, this is explained by the argument that the lifetime at 95% of the strength obtained with a normal tensile test is much longer than the time of pre-loading up to 95% of the strength. Yet, there is an alternative explanation as illustrated in Fig. 76 that leaves room for internal stresses. Two fracture envelopes are drawn for two values of the critical shear strength, viz. 0.365 and 0.375 GPa. For the fracture envelope with τ_b=0.375 GPa two tensile curves have been drawn: for a low-modulus fibre with a strength of 3.5 GPa and for a high-modulus fibre with a strength of 4.25 GPa. When the low-modulus fibre is subjected to a post-drawing step raising the modulus to a value corresponding to the drawn high-modulus curve, but at the same time introducing a residual shear stress of 0.375–0.365=0.01 GPa, no strength increase will be achieved by this process step. Suppose that in this figure σ_b^{calc} is the expected strength for a fibre with a modulus of 122 GPa and a critical shear strength of 0.375 GPa, while σ_b^{red} is the reduced strength of the fibre with the same modulus due to the residual stress. Thus, first loading up to 95% of the expected strength may increase the modulus, which should be accompanied by some increase in strength. However, this increase will not be observed after a second loading due to the introduction of residual stresses. Thus, pre-loading to 95% of the breaking strength

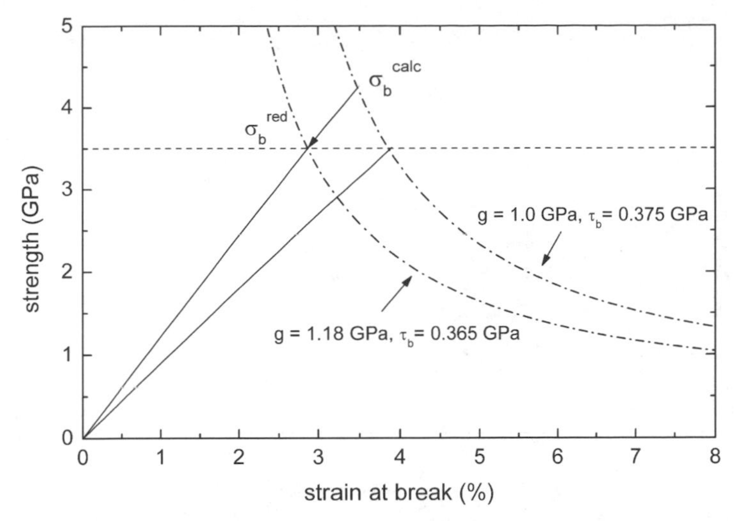

Fig. 76 Illustration of the possible effect of residual stresses on the strength of a rigid rod polymer fibre. The tensile curves of two hypothetical fibres of equal strength with a modulus of 90 and 122 GPa are schematically represented by *straight lines*. The corresponding calculated fracture envelopes have been drawn, using g=1.0 GPa for the low-modulus fibre and g=1.18 GPa for the high-modulus fibre

may result in a higher modulus, but due to the build-up of residual stresses no higher strength will be observed during a second tensile testing. Hence, because of the low relaxation rate the investigation of residual stresses in high-modulus and high-strength fibres made of rigid-rod chains deserves further attention.

As an example of the effect of large inhomogeneities on the strength and the fracture morphology, Fig. 77 shows the location of the end points of the tensile curves of filaments taken from a PpPTA yarn of low strength due to polymer degradation in the spinning process. Filaments of low strength containing degraded polymer particles showed brittle fracture morphology, whereas filaments without degraded polymer had a fibrillar fracture surface. Note that nearly all the end points approximately follow the tensile curve of the fibre. In the filaments of low strength the large particles of degraded polymer act as stress concentrators. The fibrils around these particles are loaded to very high stresses, presumably resulting in a highly oriented structure that, as we have discussed earlier, will break in a brittle way displaying a brush- or bristle-like fracture morphology.

The proposed model for creep rupture based on the condition of maximum shear strain and the Eyring reduced time model explain the observed relations concerning the lifetime of aramid, polyamide 66 and polyacrylonitrile fibres. However, with increasing temperatures, in particular above 300 °C, chemical degradation of PpPTA also determines the lifetime. Furthermore, the model

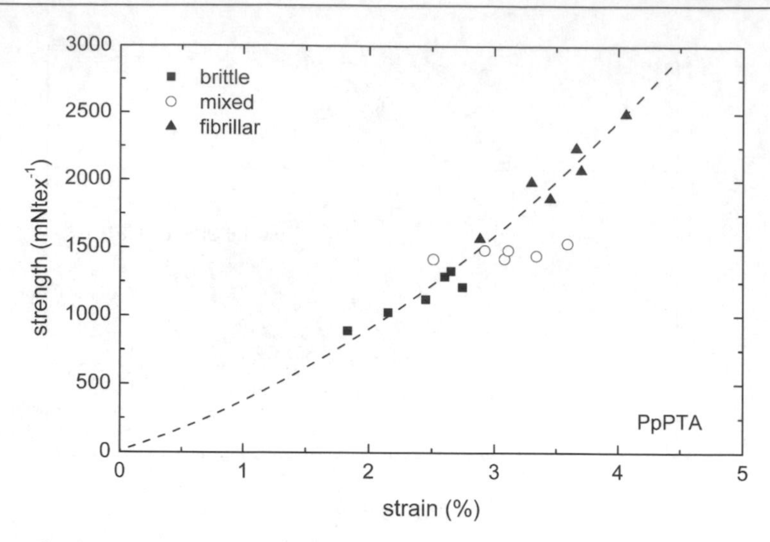

Fig. 77 The fracture morphology of strong and weak filaments taken from a single PpPTA yarn with a relatively low strength due to polymer degradation

will lose its significance if at higher temperatures the parameters, such as I_0, β, g and Ω in Eqs. 135 and 140, change their value.

Acknowledgements The authors wish to thank Dr. C.J.M. van der Heuvel for making available the load rate measurements on Twaron yarns.

References

1. Northolt MG, Baltussen JJM, Schaffers-Korff B (1995) Polymer 36:3485
2. Baltussen JJM (1996) Tensile deformation of polymer fibres. PhD thesis, Delft University of Technology
3. Baltussen JJM, Northolt MG (1996) Polym Bull 36:125
4. Baltussen JJM, Northolt MG, Van der Hout R (1997) J Rheol 41:549
5. Baltussen JJM, Northolt MG (1997) J Rheol 41:575
6. Baltussen JJM, Northolt MG (1999) Polymer 40:6113
7. Baltussen JJM, Northolt MG (2001) Polymer 42:3835
8. Northolt MG, Baltussen JJM (2002) J Appl Polym Sci 83:508
9. Baltussen JJM, Northolt MG (2003) Polymer 44:1957
10. Baltussen JJM, Northolt MG (2004) Polymer 45:1717
11. Yoon H (1990) Colloid Polym Sci 268:230
12. Termonia Y, Meakin P, Smith P (1985) Macromolecules 18:2246
13. Termonia Y, Smith P (1986) Polymer 27:1845
14. Smith P, Termonia Y (1989) Polym Commun 30:66
15. Van der Zwaag S (1989) J Test Eval 17:292
16. Penning JP, de Vries AA, van der Ven J, Pennings AJ, Hoogstraaten HW (1994) Philos Mag A69:267

17. Termonia Y (1995) J Polym Sci B33:147
18. Griffith AA (1920) Philos Trans R Soc Lond A221:163
19. Kelly A, MacMillan NH (1986) Strong solids. Clarendon, Oxford
20. Smook JH, Hamersma W, Pennings AJ (1984) J Mater Sci 19:1359
21. Van der Werff H, Akzo Nobel Research Laboratories, Arnhem, The Netherlands
22. Weibull W (1939) Ingen Vetensk Akad Handl 151:5
23. Weibull W (1951) J Appl Mech 18:293
24. Damodaran S, Desai P, Abhiraman AS (1990) J Text Inst 81:384
25. Hearle JWS, Lomas B, Cooke WD (1998) Atlas of fibre fracture and damage to textiles. Woodhead, Cambridge
26. Northolt MG, Boerstoel H, Maatman H, Huisman R, Veurink J, Elzerman H (2001) Polymer 42:8249
27. Ter Maat H, Kloos PJ, Van der Werff H, Lommerts BJ (1996) Patent EP 647282
28. Lommerts BJ (1994) Structure development in polyketone and polyalcohol fibres. PhD thesis, University of Groningen
29. Lommerts BJ, Klop EA, Aerts J (1993) J Polym Sci B31:1319
30. Wu HF, Phoenix SL, Schwartz P (1988) J Mater Sci 23:1851
31. Knoff WF (1987) J Mater Sci Lett 6:1392
32. Cook J, Gordon JE (1964) Proc Roy Soc A282:508
33. Gordon JE (1980) Structures, or why things don't fall down. Penguin, Harmondsworth
34. Hull D (1988) An introduction to composite materials. Cambridge University Press, Cambridge
35. Piggott MR (1980) Load-bearing fibre composites. Pergamon, Oxford
36. Grandbois M (1999) Science 283:1727
37. Halsey G, White HJ, Eyring H (1945) Text Res J 15:295
38. Northolt MG (1974) Eur Polym J 10:799
39. Kooijman H, Kroon-Batenburg L (1986) Internal report, University of Utrecht
40. Van der Werff H, Hofman MH, Baltussen JJM (2000) Patent EP 823499
41. Sikkema DJ (1998) Polymer 39:5981
42. Lammers M (1998) PIPD rigid-rod polymer fibres and films. PhD thesis, Swiss Federal Institute of Technology, Zurich
43. Klop EA, Lammers M (1998) Polymer 39:5987
44. Lammers M, Klop EA, Northolt MG, Sikkema D (1998) Polymer 39:5999
45. Hageman JCL, Van der Horst JW, De Groot RA (1999) Polymer 40:1313
46. Huang B, Ito M, Kanamato T (1994) Polymer 35:1329
47. Northolt MG, Sikkema DJ (1990) Adv Polym Sci 98:115
48. Picken SJ (1990) Orientational order in aramid solutions. PhD thesis, University of Utrecht
49. Picken SJ (1990) Macromolecules 23:464
50. Picken SJ, Aerts J, Doppert HL, Reuvers AJ, Northolt MG (1991) Macromolecules 24:1366
51. Picken SJ, Van der Zwaag S, Northolt MG (1992) Polymer 33:2998
52. Bovey FA, Winslow FH (1979) Macromolecules: an introduction to polymer science. Academic, New York
53. Weyland H (1980) Polym Bull 3:331
54. Coleman BD (1956) J Polym Sci 20:447
55. Zhurkov SN (1965) J Fract Mech 1:311
56. Northolt MG, Kampschreur JH, Van der Zwaag S (1989) Viscoelasticity of aramid fibres. In: Lemstra PJ, Kleintjes LA (eds) Integration of fundamental polymer science and technology. Elsevier, London, p 157
57. Northolt MG, Roos A, Kampschreur JH (1989) J Polym Sci B27:1107

58. Tobolsky A, Eyring H (1943) J Chem Phys 11:125
59. Bauwens-Crowet C, Bauwens JC, Homes G (1969) J Polym Sci A7:735
60. Strobl G (1996) The physics of polymers. Springer, Berlin Heidelberg New York
61. Kausch HH, Hsiao CC (1968) J Appl Phys 39:4915
62. Van der Werff H, Pennings AJ (1991) Colloid Polym Sci 269:747
63. Pennings AJ, Roukema M, Van der Veen A (1990) Polym Bull 23:353
64. Penning JP, Dijkstra DJ, Pennings AJ (1991) J Mater Sci 26:4721
65. Brew B, Hine PJ, Ward IM (1999) Compos Sci Technol 59:1109
66. Van der Jagt OC, Beukers A (1999) Polymer 40:1035
67. Sikkema DJ (2002) J Appl Polym Sci 83:484
68. Northolt MG, Sikkema DJ, Zegers HC, Klop EA (2002) Fire Mater 26:169
69. Northolt MG, Sikkema DJ, Patent EP 97203642
70. Kitagawa T, Murase H, Yabuki K (1998) J Polym Sci B36:39
71. Kitagawa T, Ishitobi M, Yabuki K (2000) J Polym Sci B38:1605
72. Cottrell AH (1981) The mechanical properties of matter. Krieger, Huntington
73. Woodward AE (1988) Atlas of polymer morphology. Hanser, Munich
74. Northolt MG, Veldhuizen LH, Jansen H (1991) Carbon 29:1267
75. Breedon Jones J, Barr JB, Smith RE (1980) J Mater Sci 15:2455
76. Johnson DJ (1987) Structural studies of PAN-based carbon fibers. In: Thrower PA (ed) Chemistry and physics of carbon, vol 20. Marcel Dekker, New York, p 1
77. Donnet JB, Bansal RC (1990) Mechanical properties of carbon fibers. In: Donnet JB, Bansal RC (eds) Carbon fibers. Marcel Dekker, New York, p 267
78. Northolt MG, Tabor BJ, Van Aartsen JJ (1975) Prog Colloid Polym Sci 57:225

Adv Polym Sci (2005) 178: 109–144
DOI 10.1007/b104208

Advances in Inorganic Fibers

Toshihiro Ishikawa (✉)

Ube Research Laboratory, Ube Industries Ltd., 1978-5, Kogushi, Ube City,
Yamaguchi Prefecture, 755-8633, Japan
24613u@ube-ind.co.jp

Abstract Many types of ceramic fibers have been developed over the last 60 years. Of these, carbon fiber has grown into a large market involving fiber-reinforced plastics and light-weight heat-resistant composites. In addition, many modifications of SiC-based fibers have been achieved. SiC-polycrystalline fibers (Hi-Nicalon Type S and SA fiber) with excellent heat-resistance have been developed and their many applications have been examined. A tough, thermally conductive SiC-based ceramic (SA-Tyrannohex) composed of a highly ordered, close-packed structure of hexagonal columnar fibers has also been produced. Recently, using a precursor polymer containing low molecular weight additives, which can be converted into an objective functional material, excellent functional ceramic fibers with gradient surface layers were also developed. In this chapter, representative properties of these fibrous materials and their expected applications are described.

Keywords Inorganic fiber · Oxide fiber · Silicon carbide fiber · Heat resistance · Photocatalyst

1 Introduction

Regarding inorganic fibers, glass fibers and carbon fibers are well known all over the world. In addition, alumina-silica fibers, single crystalline oxide fibers and silicon carbide fibers, which show excellent oxidation resistance at high temperatures in air, have been developed and commercialized except for single crystalline oxide fibers. Of these, glass fibers and carbon fibers have made a lot of progress in the field of reinforced plastics. In particular, carbon fiber has already established a very big market. Oxide fibers such as alumina-silica fiber are used for insulating fabric materials of space shuttles , and so forth. However, at present those oxide fibers have not been able to make such a large market as have carbon fibers whereas research on composite materials using oxide interfaces are continuing in many research laboratories [1]. On the other hand, silicon carbide fibers have achieved great progress in specific characteristics [2–5] over the last six years. As a result, lots of advanced research on ceramic composite materials to make the best use of their excellent heat-resistance and oxidation-resistance have been carried out [6]. Furthermore, an excellent functional ceramic fiber with a gradient surface layer was recently synthesized from an organosilicon polymer making full use of the technology of the aforementioned fiber production. In this case, a general process for in-situ formation of functional surface on ceramic fibers was proposed [7]. This process is characterized by controlled phase separation ("bleed out") of additives, analogous to the normally undesirable outward loss of low-molecular-mass components from some plastics; subsequent calcination stabilizes the compositionally changed surface region, generating a functional surface layer. This approach is applicable to a wide range of materials and morphologies, and should find use in catalysts (for example, photocatalysts), composites and environmental barrier coatings.

The objective of this chapter is to review the advance in these fibers and refer to the prospect for the future technology of inorganic fibers.

2 Carbon Fiber and its Composites

Commercial production of carbon fiber was started in 1970, the second longest history only after glass fiber. Several Japanese companies monopolize the manufacturing and marketing of carbon fibers all over the world. Originally, carbon fiber was used in fiber-reinforced plastics (FRP) in structural materials, making the best use of the light weight and high strength. The first application to aerospace materials was the use of PAN-based carbon fiber (TORAY, T-300) in 1976, when an energy-saving program for manufacturing airplanes was started in the USA. After that, active market development led to use in the tail assembly unit of a passenger plane (Boeing 777), in which a prepreg sheet

Fig. 1 The outside appearance of the Space Shuttle

(TORAY, P2301-19) composed of a high strength carbon fiber (TORAY, T-800) and a high strength epoxy resin (3900-2) was formally certified.

Though the industrial position of carbon fiber was founded mainly on the aforementioned FRPs, other applications making good use of the light weight, high strength and heat resistance have been investigated for many years. The temperature of the outside wall of the Space Shuttle (Fig. 1) locally increases to over 1250 °C by air friction while passing through the atmosphere. In order to withstand the high temperatures, many types of thermal protection system (TPS) are adopted for the surface of the Space Shuttle. Of these, for several parts (nose and front side of the wing of the shuttle) heating up to the highest temperatures (over 1250 °C), black tiles constructed from carbon fiber/carbon composites (C/C composites) are used.

In NASA Glenn Research Center, research on carbon fiber-reinforced SiC composites (C/SiC) is actively being continued to produce thermostructural materials which can be used in a much more severe environment [8]. The average mechanical properties of C/SiC are shown in Table 1 and the typical stress-strain curves of the composites up to high temperatures are shown in Fig. 2. As can be seen from this figure, the C/SiC maintained about 74% of the room temperature strength up to 1480 °C. At present, this type of thermostructural material is under development for the production of engine parts. On the other hand, in France, to produce the abrasion material for a rocket nozzle, a low cost DCCVI process (Direct Cooling Chemical Vapor Infiltration) was proposed instead of the former ICVI process (Isothermal

Table 1 Typical tensile properties of isothermal chemical vapor infiltration processed C/SiC

Composite architecture	Tensile modulus GPa	Tensile strength MPa
[0/90]	83	434
[0/±60]	69	269

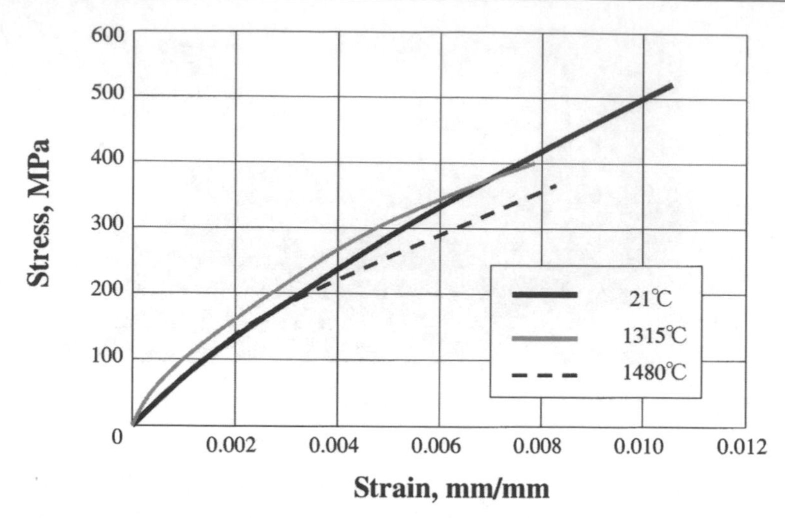

Fig. 2 The typical tensile-stress/strain curves of the [0/90] plain weave C/SiC composites up to high temperatures

Chemical Vapor Infiltration) [9]. As the carbon fiber itself is the main material, a low cost production process for the composite itself and for coating the composite might be advantageously investigated.

3 Oxide Fibers and their Composites

Except for glass fibers, synthetic oxide fibers (alumina-silica fibers) were produced in the early 1970s. These fibers, under the trade name Saffil (ICI), today represent the most widely used filamentary reinforcement for light alloys. Amorphous continuous fibers, based on mullite with boron added, were also produced around the same time by 3 M under the trade name of Nextel-312. Small-diameter continuous alpha-alumina fibers were produced, first by Du Pont, in the 1970s and these began to be incorporated into metal matrix composites toward the end of that decade. The composites that were produced showed great improvements in stiffness and creep resistance when compared to unreinforced aluminum; however, high cost and the brittleness of the fibers then limited their use. Since 1980, other fibers based on alumina, often containing a small amount of silica, or mullite have been produced which are easier to handle. However, often the presence of silica results in the reduction of Young's modulus and creep resistance.

Oxide fibers such as silica and alumina, which have oxidation stability and insulating properties, are used for heat insulators, such as the TPS (Fig. 3) on the upper part of the Space Shuttle. On the surface of the TPS, a coating of sil-

Fig. 3 Silica/alumina-based blanket for the TPS on the upper part of the Space Shuttle

ica-based oxide material is applied to produce the stiffness to withstand the aerodynamic stress. As the upper limit of the usable temperature of the present silica-based coating material is 650 °C, above this temperature the molten coating material unites with the fibrous reinforcement to show the brittle behavior. Thus, regarding the TPS, research to increase the hear-resistance of the interface and coating materials, not to mention research on the fiber itself, has been performed [10].

Under these conditions, many types of continuous oxide fiber were developed. The physical properties of these oxide fibers are shown in Table 2 [11]. Methods for preparation of these oxide fibers include spinning of a sol, a solution, or slurry, usually containing fugitive organics as part of a precursor.

Table 2 Physical properties of various oxide fibers

Fiber	Manufacturer	Composition, wt%	Young's modulus E, GPa	Diameter μm	Density $g \cdot cm^{-3}$
Almax	Mitui Mining	α-Al_2O_3	320~340	10	~3.6
Altex	Sumitomo	$15SiO_2$-$85Al_2O_3$	200~230	9~17	~3.2
Fiber FP	Du Pont	α-Al_2O_3	380~400	~20	3.9
Nextel 312	3 M	24SiO2-14B2O3-62Al2O3	150	10~12	2.7~2.9
Nextel 440	3 M	$28SiO_2$-$2B_2O_3$-$70Al_2O_3$	220	10~12	3.05
Nextel 480	3 M	$28SiO_2$-$2B_2O_3$-$70Al_2O_3$	220	10~12	3.05
Nextel 550	3 M	$27SiO_2$-$73Al_2O_3$	193	10~12	3.03
Nextel 610	3 M	Al_2O_3	373	10~12	3.75
Nextel 720	3 M	$15SiO_2$-$85Al_2O_3$	260	10~12	3.4
Nextel Z-11	3 M	$32ZrO_2$-$68Al_2O_3$	76	10~12	3.7
PRD-166	Du Pont	$80Al_2O_3$-$20ZrO_2$	360–390	14	4.2
Saftil	ICI	$4SiO_2$-$96Al_2O_3$	100	~20	2.3
Safikon	Safikon	Al_2O_3	386~435	3	3.97
Sumica	Sumitomo Chemical	$15SiO_2$-$85Al_2O_3$	250	75~225	3.2

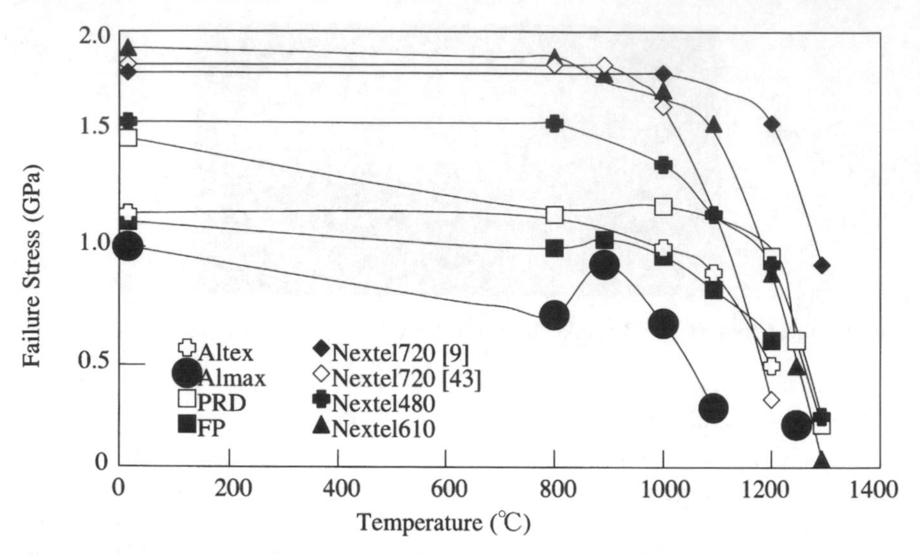

Fig. 4 The high temperature strengths of representative oxide fibers including the Nextel fibers (from: Bunsell AR, Berger M-H. Fine ceramic fibers. Marcel Dekker, New York)

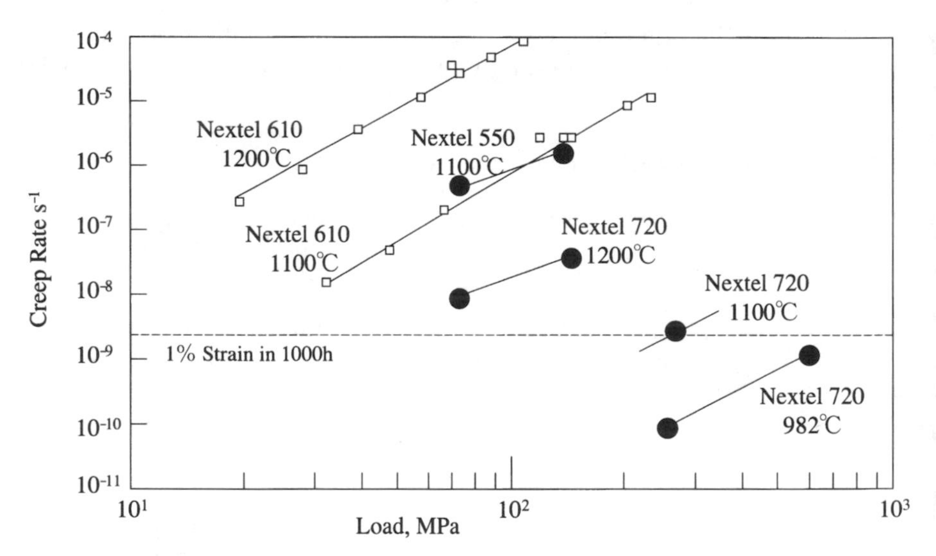

Fig. 5 Comparison of creep behavior of three different Nextel fibers: note superior creep resistance of Nextel 720 fiber (data from 3 M Co.)

Among the oxide fibers, alumina-based and aluminosilicate fibers with near mullite compositions are those most widely used. In particular, the 3 M Company has developed a series of oxide fibers. This series, called the Nextel fibers, mainly consists of alumina and mullite type fibers. Nextel ceramic oxide fibers are typically transparent, non-porous, and have a diameter of 10–12 μm. In this series, Nextel 550 is a mullite fiber. Nextel 610 fiber, a polycrystalline alpha-alumina fiber, has the highest modulus in this series, while Nextel 720, an alumina+mullite fiber, has the highest temperature- and creep-resistance in the group. The high temperature strength and the creep behavior of representative oxide fibers including these Nextel fibers are shown in Figs 4 and 5, respectively. On the other hand, in order to increase in the high temperature strength of the oxide fiber, lots of research on single crystalline oxide fibers was performed at many companies in the 1970s. However, as can be seen in Fig. 6, all of these single crystalline oxide fibers show a remarkable decrease in the tensile strengths over 1100 °C. For the purpose of achieving an increase in the high temperature strength, a eutectic fiber consisting of interpenetrating phases of alpha-alumina and yttrium-aluminum-garnet (YAG) was developed. The structure depends on the conditions of manufacture, in particular the drawing speed, but can be lamellar and oriented parallel to the fiber axis. This fiber showed superior creep-resistance up to very high temperatures compared with other types of oxide fibers (Fig. 7). At present, this fiber has not been commercialized because of the large fiber diameter (50–150 μm) and fewer production facilities.

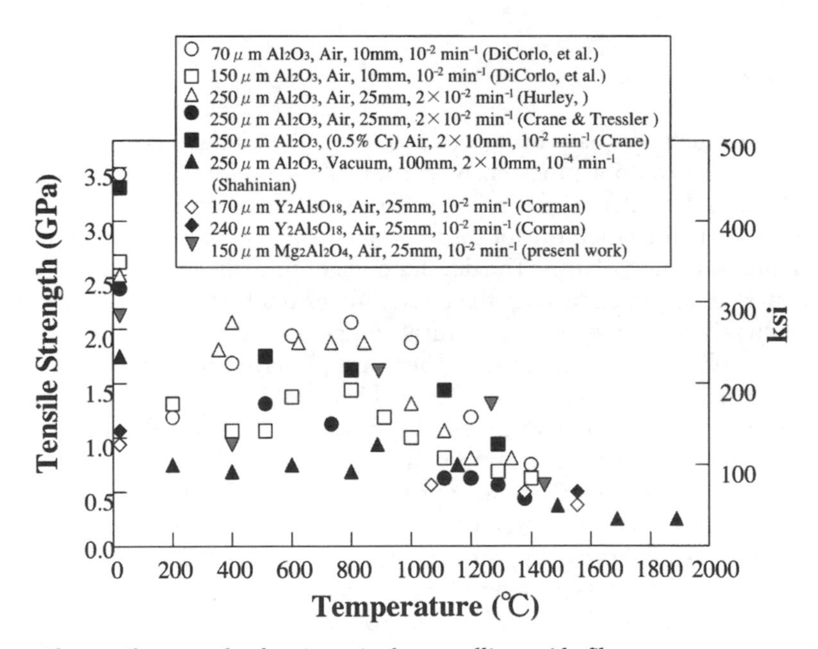

Fig. 6 The tensile strength of various single crystalline oxide fibers at room temperature and elevated temperatures

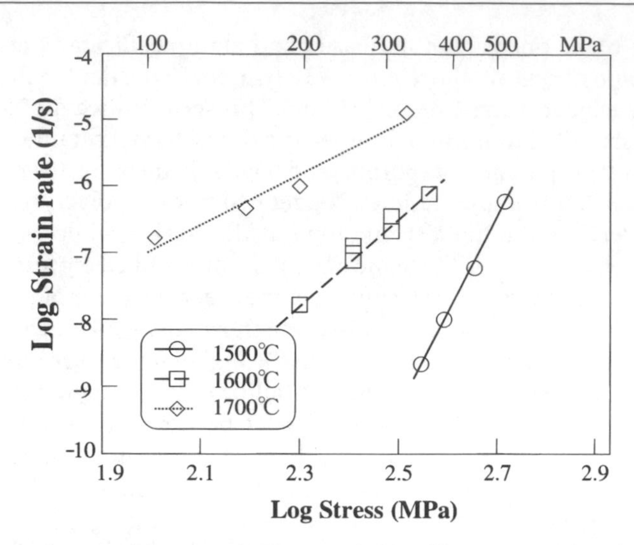

Fig. 7 Creep-resistance of the eutectic fiber consisting of interpenetrating phases of alpha-alumina and yttrium-aluminum-garnet (YAG) up to very high temperatures compared with other types of oxide fibers

4 Silicon Carbide Fibers

4.1 Overview

Silicon carbide fibers have been produced since the mid-1960s by chemical vapor deposition onto a tungsten or carbon filament core. Large filaments of 100–140 μm diameter were produced: however, their dimensions and the lack of flexibility limited their use for the reinforcement of metals such as aluminum, titanium, and intermetallics. The development of fine silicon carbide fibers with diameters of 10 μm opened up the possibility of reinforcing ceramic materials to produce high-temperature structural composites. This was at the beginning of the 1980s. Fine silicon carbide fibers are prepared by the melt spinning, crosslinking, and pyrolysis of an organosilicon polymer (Fig. 8). They primarily consist of fine SiC grains and several phases, that considerably influence the characteristic behavior of the fibers. Since the production of the first Nicalon fibers in 1980 by Nippon Carbon, improvements of the fabrication route and/or modifications of the precursor polymer have permitted the development of other SiC-based fibers by Nippon Carbon and Ube Industries. Lots of excellent SiC-based fibers such as Nicalon fibers (Nicalon, Hi-Nicalon, Hi-Nicalon Type S) of Nippon Carbon, Tyranno fibers (LoxM, ZMI, ZE, SA fiber) of Ube Industries, and Sylramic of Dow Corning were developed [2, 12–16]. The physical properties of these SiC-based fibers are shown in Table 3. All of these fibers

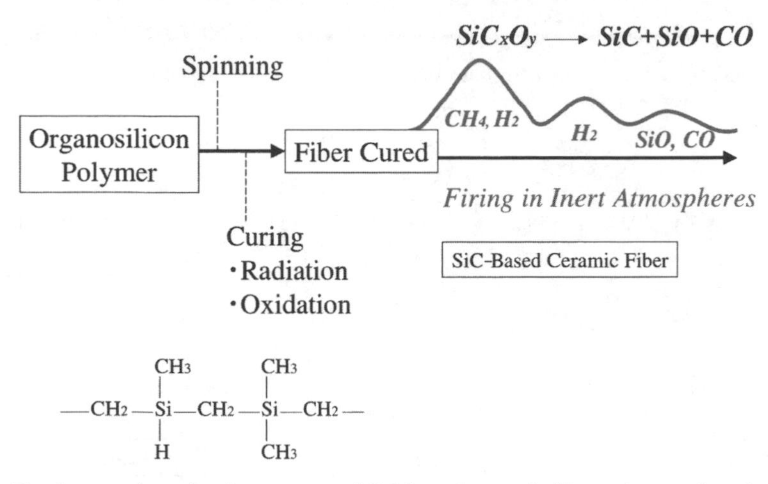

Fig. 8 Fundamental production process of SiC-based ceramic fiber using a polycarbosilane

Table 3 Physical properties of representative SiC-based fibers

(shaded): *Nearly stoichiometric SiC fiber*

	SiC Fibers							
	Nicalon			Tyranno				Sylramic
	NL-200	Hi-Nicalon	Hi-Nicalon–s	Lox M	ZMI	ZE	SA*	
Atomic Composition	$SiC_{1.34}O_{0.36}$	$SiC_{1.39}O_{0.01}$	$SiC_{1.05}$	$SiTi_{0.02}C_{1.37}O_{0.32}$	$SiZr_{<0.01}C_{1.44}O_{0.24}$	$SiZr_{<0.01}C_{1.52}O_{0.05}$	SiC O,Al$_{<0.008}$	$SiCTi_{0.02}B_{0.09}O_{0.02}$
Tensile Strength (GPa)	3.0	2.8	2.6	3.3	3.4	3.5	2.8	3.0
Tensile Modulu (GPa)	220	270	410	187	200	233	410	420
Elongation (%)	1.4	1.0	0.6	1.8	1.7	1.5	0.7	0.7
Density ($g \cdot cm^{-3}$)	2.55	2.74	3.10	2.48	2.48	2.55	3.02	>3.1
Diameter (μm)	14	14	12	8 & 11	8 & 11	11	8 & 10	10
Specific Resistivity (Ω · cm)	10^3-10^4	1.4	0.1	30	2.0	0.3	—	—
Thermal Expansion coeff. (10^{-6}/K)	3.2 (25-500℃)	3.5 (25-500℃)	—	3.1	4.0	—	4.5 (20-1320℃)	—
Thermal Conductivity (W/mK)	2.97(25℃) 2.20(500℃)	7.77(25℃) 10.1(500℃)	18.4 (25°C) 16.3 (500°C)	—	2.52	—	64.6	40·45

* T.Ishikawa, et al., *Nature*, 391, 773-775 (1998)

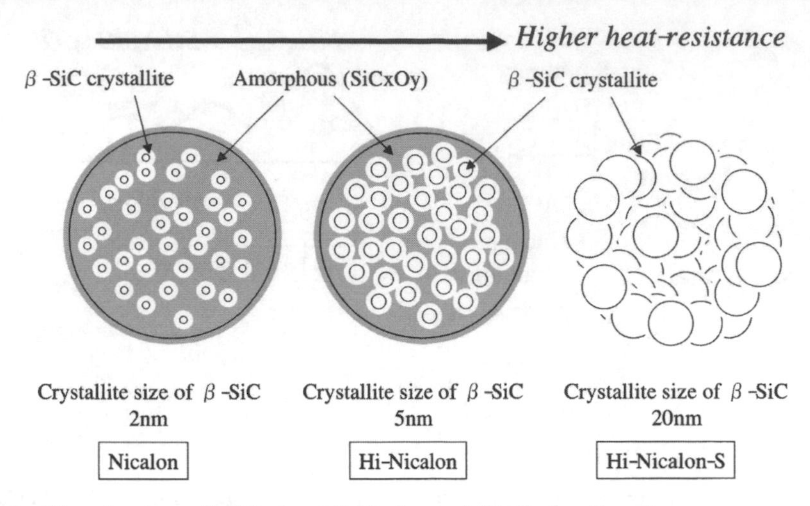

Fig. 9 Microstructures for Nicalon, Hi-Nicalon and Hi-Nicalon Type S

have high strength and high heat-resistance, and applications as reinforcement for thermostructural ceramic composites are actively being developed. Microstructures for Nicalon, Hi-Nicalon and Hi-Nicalon Type S are shown in Fig. 9.

As can be seen from this figure, the heat-resistance was remarkably improved by the drastic changes in the microstructure from amorphous to polycrystalline structure. Another type of SiC-based fiber, SA fiber (2), has a sintered SiC polycrystalline structure and includes very small amounts of aluminum. This fiber exhibits outstanding high temperature strength, coupled with much improved thermal conductivity and thermal stability compared with the Nicalon and Hi-Nicalon fibers. The fabrication cost of the SA fiber is also reduced to near half of that of the Hi-Nicalon Type S [17]. The SA fiber makes SiC/SiC composites even more attractive to the many applications [18]. In the next section, the production process, microstructure and physical properties of the SA fiber are explained in detail.

4.2 Excellent Heat-Resistant SiC-Based Fiber (SA Fiber)

The high-temperature stability of SiC-based ceramics is well-known, and therefore its composite materials have been investigated for application to high-temperature structural materials [19–21]. However, well-known SiC-based fibers and matrix-materials stained with alkali salt are easily oxidized at high temperatures in air [22]. This would be a serious problem when these materials are used near the ocean or in a combustion gas containing alkali elements. In particular, a silicon carbide fiber containing boron (a well-known sintering aid for SiC) over 1 wt% was extensively oxidized under the above condition. In this section it is explained that a sintered SiC fiber (SA fiber) containing a very small amount of Al showed high-strength, high-modulus, excellent high temperature stability and prominent alkali-resistance. Moreover, this fiber shows excellent creep resistance at very high temperatures compared with well-known SiC-based fibers. The SA fiber was synthesized at a very high temperature over 1800 °C using an amorphous Si-Al-C-O fiber as the starting material. Its high strength of over 2.5 GPa was maintained up to 2000 °C in Ar atmosphere and very little weight loss (only 1.8 wt%) was observed up to 2200 °C. The fabrication process of the SA fiber is as follows.

Si-Al-C-O fiber (an intermediate fiber) was synthesized by the use of polyaluminocarbosilane prepared by the reaction of polycarbosilane $(\text{-SiH(CH}_3\text{)-CH}_2\text{-})_n$ with aluminum(III)acetylacetonate. The reaction of polycarbosilane with aluminum(III)acetylacetonate proceeded at 300 °C in a nitrogen atmosphere by the condensation reaction of Si-H bonds in polycarbosilane and the ligands of aluminum(III)acetylacetonate accompanied by the evolution of acetylacetone [23], and then the molecular weight increased by the cross-linking reaction with the formation of a Si-Al-Si bond. Polyaluminocarbosilane was melt-spun at 220 °C, and then the spun fiber was cured in air at 160 °C. The cured fiber was continuously fired in inert gas up to 1300 °C to obtain an amorphous Si-Al-C-O fiber (an intermediate fiber). This fiber contained non-stoichiometric excess carbon and oxygen of about 12 wt%. The Si-Al-C-O fiber was converted into the SA fiber by way of decomposition accompanied by the release of CO gas at temperatures from 1500 °C to 1700 °C and sintering at temperatures over 1800 °C [2]. In this sintering process, aluminum plays a very important role

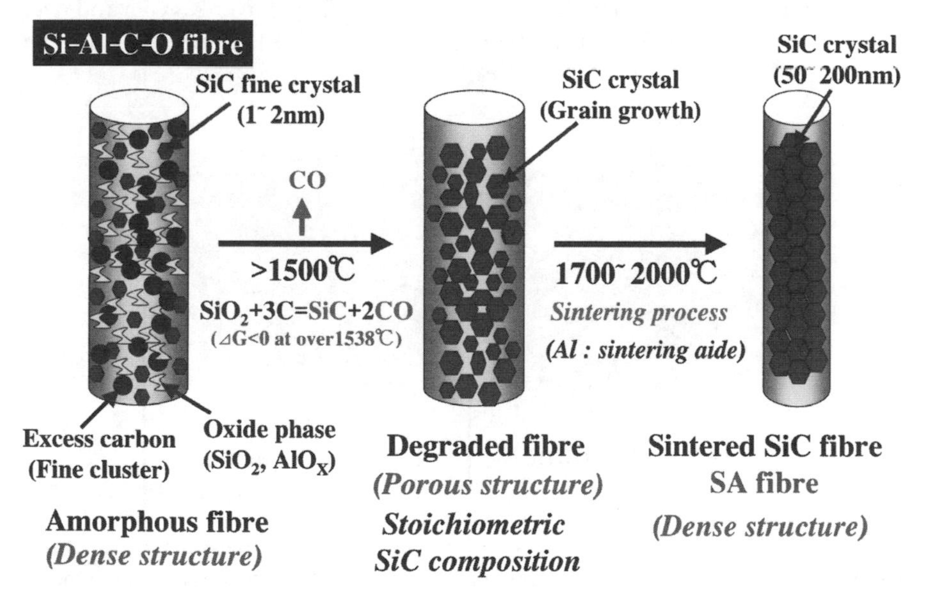

Fig. 10 Structural transition from Si-Al-C-O fiber to SA fiber at high temperatures

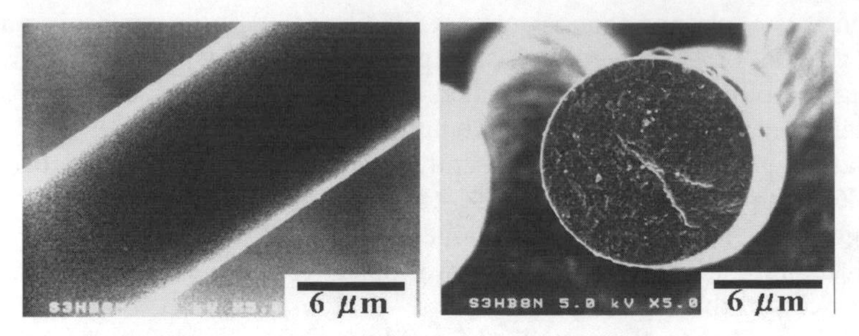

Fig. 11 Surface and cross-section of the SA fiber

as a sintering aid. This conversion process is schematically shown in Fig. 10. In order to obtain a strong SA fiber, the content of aluminum in the fiber has to be controlled to less than 1 wt%. The SA fiber with a controlled amount of aluminum under 1 wt% showed a smooth surface and densified structure (Fig. 11). Moreover, in this case, SA fiber showed transcrystalline fracture behavior (Fig. 12B). On the other hand, the undesirable sintered fiber with a large amount of aluminum showed intercrystalline fracture behavior (Fig. 12A). These phenomena are presumed to be related to the upper limit of solid-soluble concentration of aluminum to SiC crystal [24]. From the TEM image (Fig. 13) of the desirable SA fiber, no obvious second phase is observed at the grain boundary and triple point. EDS (energy dispersive X-ray spectroscopy) spectra taken at these places did not indicate the presence of aluminum within the detectability limit of aluminum (~0.5 wt%) for the EDS system used.

SA fiber showed high tensile strength and modulus of over 2.5 GPa and over 300 GPa, respectively. The initial strength was preserved after heat-treatment

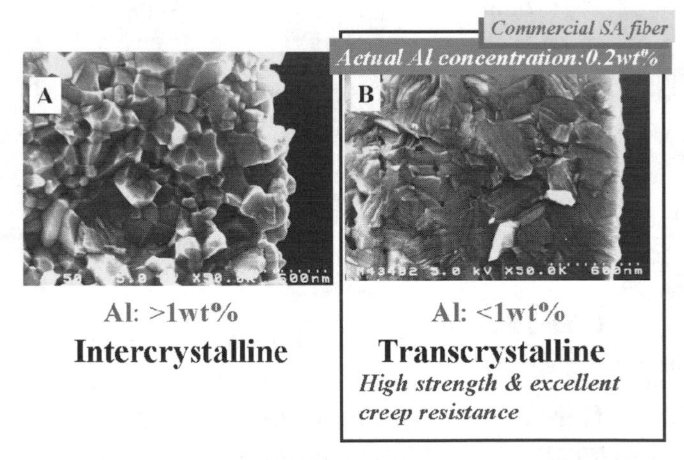

Fig. 12A,B The relation between the aluminum content and the fracture behavior of the SA fiber

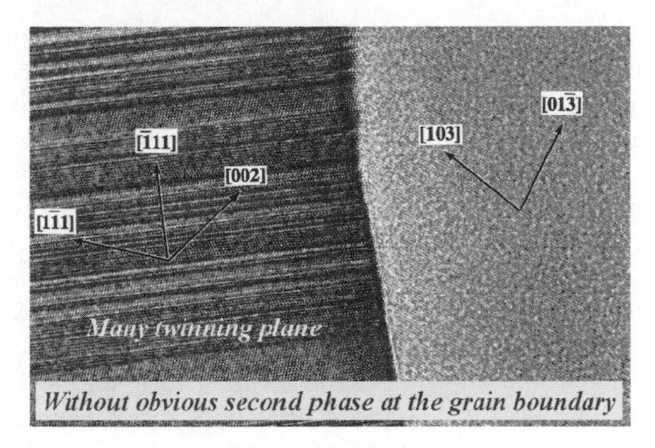

Fig. 13 TEM image at the grain boundary of SA fiber

at 2000 °C for 1 h in Ar (Fig. 14). On the other hand, the strength of a representative SiC fiber (Hi-Nicalon) was reduced to 65% and 40% of the initial strength by heating in Ar for 1 h at 1550 °C and 1800 °C, respectively. And furthermore, the relative strength of SA fiber exposed in air at 1000 °C or 1300 °C for 100 h was 100% or 55%, respectively, whereas that of Hi-Nicalon exposed under the same testing condition was 73% or 23%, respectively. From these results, this fiber is found to show excellent heat-resistance and oxidation-resistance among all types of SiC-based continuous fibers. This thermal stability of SA fiber is assumed to be caused by the densified and sintered structure composed of nearly stoichiometric SiC crystal.

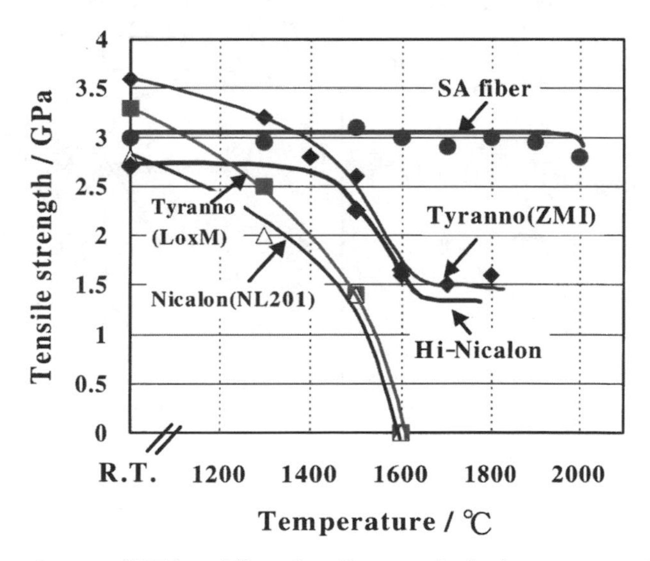

Fig. 14 Heat resistance of SiC-based fibers (tensile strength after heat-treatment in Ar for 1 h)

As mentioned above, this type of SiC-based fiber has been mainly developed for high temperature application. Therefore, its heat-resistance, creep-resistance and chemical resistance under severe conditions at high temperatures are very important. Figure 15 shows the result of the creep-resistance of SA fiber in comparison with other SiC-based fibers. In this testing, the tensile creep behaviors of four types of SiC-based fibers at 1300 °C and 1400 °C were evaluated. In this figure, SA1 and SA3 fibers are shown. The differences between SA1 and SA3 are shown in Fig. 16. As can be seen from Fig. 15, the order of the creep resistance of these fibers is as follows: Hi-Nicalon Type S>SA3>Sylramic>SA1. All of these crystalline fibers are found to show excellent creep resistance. Furthermore, the production cost of the SA fiber (SA3 and SA1) is the lowest of these fibers, because SA fiber does not adopt the following two types of production process. One is the electron curing process, which is an important process for preparing Hi-Nicalon Type S. In the case of SA fiber, low cost curing process using air at relatively low temperatures is adopted. This process is almost similar to an oxidation process. Second is the chemical vapor incorporation of a sintering aid into the intermediate inorganic fiber. This process is very important for preparing Sylramic, but at very high cost and lower production ability. In the case of SA fiber, an element as the sintering aid (aluminum) is introduced into the precursor polymer, and then the aforementioned second process like or such as preparing Sylramic (chemical vapor incorporation of a sintering aid) is not needed.

Next, the alkali-resistance of SA fiber is described in detail compared with other type of SiC polycrystalline fiber using boron instead of aluminum as the

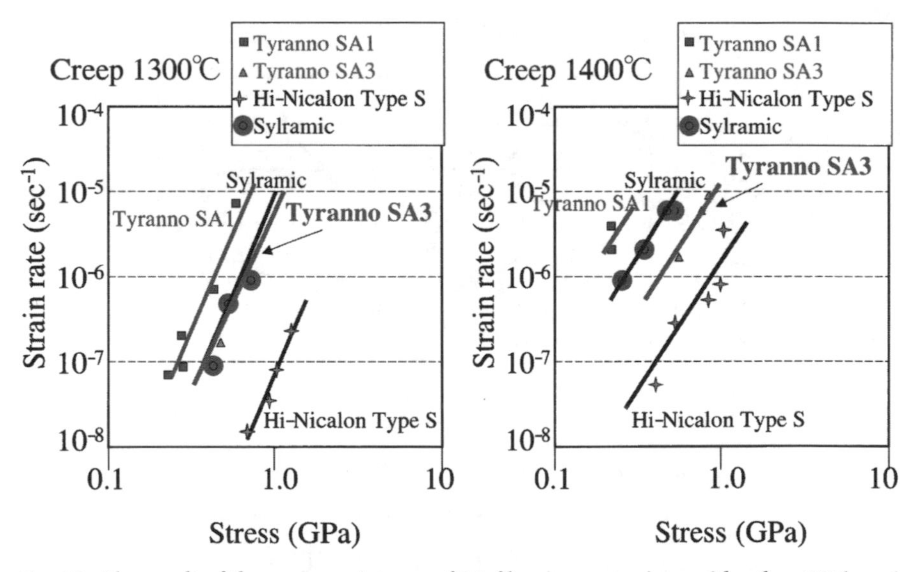

Fig. 15 The result of the creep-resistance of SA fiber in comparison with other SiC-based fibers

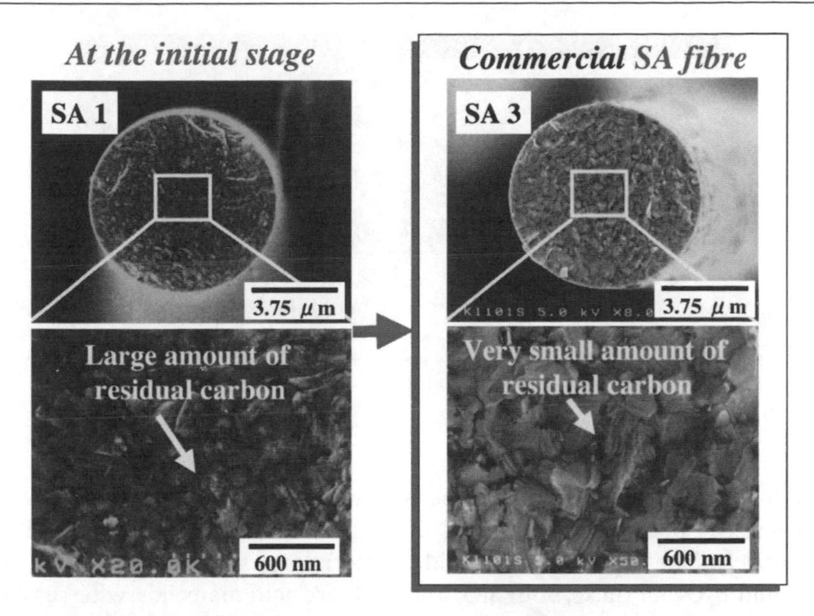

Fig. 16 The differences between SA1 and SA3

sintering aid. SiC ceramics have been well-known as very stable materials at high temperatures in air. The good oxidation resistance of SiC in air is due to the protection provided by the layer of vitreous silica (oxidation product). However, in the case of ordinary SiC materials, the silica can devitrify rapidly in the presence of alkali elements, which results in enhanced oxidation. When these SiC materials are used near the ocean or in a combustion gas containing alkali elements, these phenomena cause serious problems. Therefore, we examined the stability of SA fiber in the presence of alkali salt. To study the effect of salt existence on the fiber at high temperatures in air, the following experiments were conducted [22]. The SA fiber was immersed in deionized water saturated with NaCl at room temperature for 15 min, and then annealed at 1000 °C for 2 h in air. At the same time, in order to conduct a comparative study, a SiC crystalline fiber including boron of 1.5 wt%, prepared by heat-treating an amorphous Si-C-O fiber under an argon gas atmosphere including B_2O_3 vapor, was also tested in the same way. Figure 17 shows an FE-SEM micrograph of the surfaces of the tested fibers. As can be seen from this micrograph, the SA fiber including a small amount of aluminum exhibits a very smooth surface. On the other hand, the SiC crystalline fiber including boron of 1.5 wt% instead of aluminum was extensively oxidized, and many cracks could then be observed on the surface. Both aluminum and boron with small amount of carbon are well-known as good sintering aids for SiC crystal [25, 26]. However, from the result of the above-mentioned test, aluminum is found to be suitable for the synthesis of alkali resistant SA fiber. The difference in the oxidation behavior between the two fibers is presumed to be related to the basicity of the oxide ma-

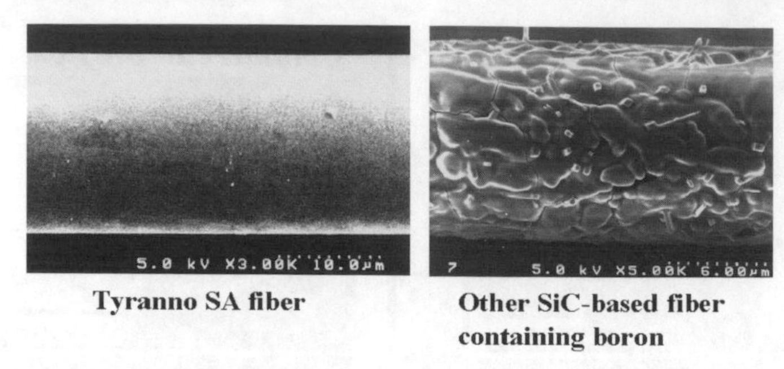

Fig. 17 The surface of the tested fibers as follows (fibers were immersed in deionized water saturated with NaCl at room temperature for 15 min, and then annealed at 1000 °C for 2 h in air)

terial formed by oxidation of the fiber's element. The main elements of the SA fiber and the boron-containing SiC crystalline fiber are Si-C-Al and Si-C-B, respectively. The representative oxide materials of these elements are SiO_2, Al_2O_3 and B_2O_3. Of these, both SiO_2 and B_2O_3 are acid materials, whereas Al_2O_3 is an amphoteric material. That is to say, the surface oxide layer of the oxidized former fiber (SA fiber) includes an amphoteric material. Generally, amphoteric or basic oxide-materials show good alkali-resistance compared with acid oxide-materials [27]. In addition, in the comparative study, the strength of a representative SiC fiber (Hi-Nicalon), which does not include both boron and aluminum, was enormously reduced to below 10% of the initial strength under the above testing condition. It is concluded that the inclusion of a small amount

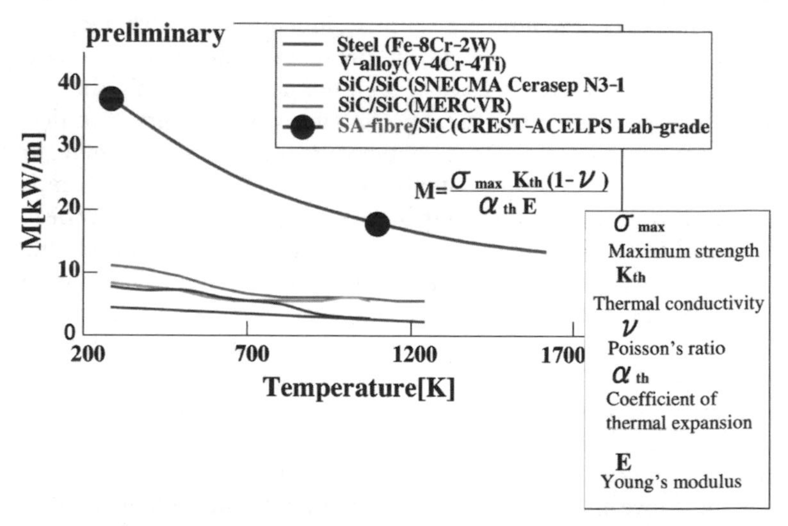

Fig. 18 Coefficient of fracture-resistance to thermal stress of several materials (Institute of Advanced Energy, Kyoto University, Japan Science and Technology Corporation)

Fig. 19 8HS fabric made of a very thin SA fiber (diameter: 7~8 μm)

of aluminum in the SA fiber plays a very important role in obtaining very excellent oxidation-resistance in the presence of an alkali element.

Finally, excellent thermal conductivity of SA fiber (about 64 W/mK), which is twice as high as that of alumina, is introduced. SA fiber can be used for thermal shock resistant composites making the best use of the high thermal conductivity. The thermal-shock resistant ability of the SA fiber-reinforced composite is shown in Fig. 18 compared with other composites. As can be seen from the equation shown in the figure, the larger the maximum strength and thermal conductivity, the higher the index number for realizing the excellent thermal shock-resistance. It should be noted that the fracture-resistance to thermal stress remarkably increased by the use of SA fiber. In order to fabricate the good thermo-structural composites, SA fiber is mainly used as the fabric shown in Fig. 19.

4.3 Thermally Conductive, Tough Ceramic Making the Best Use of Production Process of SA Fiber

A new type of toughened SiC-based material containing perfectly close-packed, very fine hexagonal-columnar fibers, which consist of a sintered structure of beta-SiC crystals, is introduced. At the interface between the hexagonal-columnar fibers, a very thin interfacial carbon layer uniformly exists, which results in a fibrous fracture behavior. This material has a very high fiber volume fraction (~100%), and showed excellent oxidation resistance even at 1600 °C in air and maintained the initial high strength up to such high temperatures. Furthermore, this material (SA-Tyrannohex) showed relatively high thermal conductivity even at high temperatures (>1000 °C), which would allow its use in the fabrication of high-temperature heat exchanger components. First, the production process is shown as follows.

An amorphous Si-Al-C-O fiber, which is the intermediate fiber for preparing the aforementioned SA fiber and also the starting material for fabricating the SA-Tyrannohex, was synthesized from polyaluminocarbosilane, which was prepared by the reaction of polycarbosilane $(\text{-SiH(CH}_3\text{)-CH}_2\text{-})_n$ with aluminum(III)acetylacetonate. The reaction of polycarbosilane with aluminum-(III)acetylacetonate proceeded at 300 °C in a nitrogen atmosphere through the condensation reaction of Si-H bonds in polycarbosilane and the ligands of aluminum(III)acetylacetonate accompanied by the evolution of acetylacetone [23]. The molecular weight then increased due to the cross-linking reaction in the formation of Si-Al-Si bond. Polyaluminocarbosilane was melt-spun at 220 °C, and then the spun fiber was cured in air at 153 °C. The cured fiber was continuously fired in inert gas up to 1350 °C to obtain an amorphous Si-Al-C-O fiber with diameters of about 10 μm (87% between 8 and 12 μm). This fiber contained a nonstoichiometric amount of excess carbon and oxygen (about 11 wt%). Unidirectional sheets with thicknesses of about 100 μm were prepared with the Si-Al-C-O fiber. Laminated materials, prepared with the unidirectional sheets, were hot-pressed at 1800 °C and 50 MPa to obtain the SA-Tyrannohex mainly composed of beta-SiC crystals [28]. During hot-pressing, the amorphous Si-Al-C-O fiber was converted into a sintered SiC fiber by way of decomposition, which released CO gas, and a sintering process accompanied by a morphological change from a round columnar shape to a hexagonal columnar shape. In this sintering process, the concentration of aluminum in the

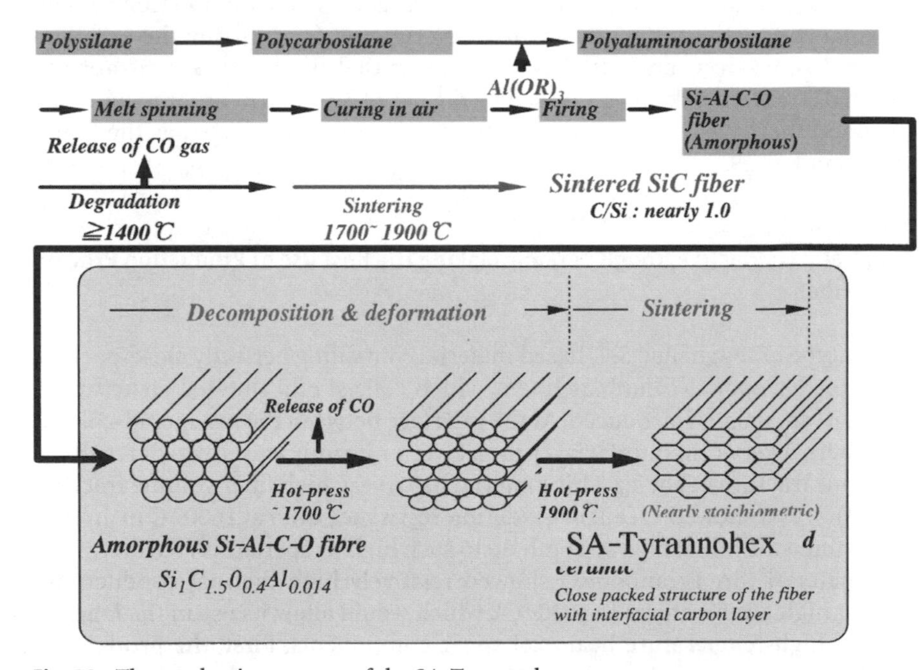

Fig. 20 The production process of the SA-Tyrannohex

fiber has to be controlled to less than 1 wt%. Such a sintered SiC fiber element (with <1 wt% Al) showed a densified structure and transcrystalline fracture behavior, which is almost same as that of the aforementioned SA fiber. The production process of the SA-Tyrannohex is shown in Fig. 20.

The SA-Tyrannohex showed a perfectly close-packed structure of the hexagonal-columnar fibers with a very thin interfacial carbon layer, as can be seen in Fig. 21A. The interior of the fiber element was composed of sintered beta-SiC crystal without an obvious second phase at the grain boundary and triple points. Energy-dispersive X-ray (EDX) spectra taken at these places did not indicate the presence of aluminum within the detectability limit (~0.5 wt%) for the EDX system used. Because of the existence of the very thin interfacial carbon layer, the SA-Tyrannohex exhibited a fibrous fracture behavior and a large amount of fiber pull-out could be observed, as shown in Fig. 21B. Accordingly, the SA-Tyrannohex showed nonlinear fracture behavior and relatively high fracture energy (1200 J/m^2) compared with monolithic ceramic (for example, 80 J/m^2 of silicon nitride) (Fig. 22). This is closely related to the high fiber volume fraction and the existence of a strictly controlled interphase. The interfacial carbon layer has a turbostratic layered structure oriented parallel to the fiber surface.

The SA-Tyrannohex showed excellent high-temperature properties compared to ordinary SiC-CMCs. The result of a four-point bending test of the SA-Tyrannohex up to high temperatures is shown in Fig. 23. The SA-Tyrannohex retained its initial strength up to 1600 °C, whereas the Hi-Nicalon SiC/SiC showed a definite decrease in strength at temperatures above 1200 °C [29]. Other types of SiC/SiC composites also show the same behavior as the Hi-Nicalon SiC/SiC [30, 31]. In general, the high-temperature properties of conventional SiC-CMCs are closely related to the high-temperature strength of the reinforcing fiber. The strength of Hi-Nicalon gradually decreases with an increase of measuring temperature, even in an inert atmosphere; at 1500 °C the strength is about 43% of its low-temperature strength [32]. Accordingly, it has been concluded that the above reduction in the strength of the Hi-Nicalon

Fig. 21 The cross-section and fracture surface of SA-Tyrannohex. This shows relatively large fracture energy (~2000 J/m^2), higher proportional limit (about 120 MPa) and high tensile strength (200 MPa)

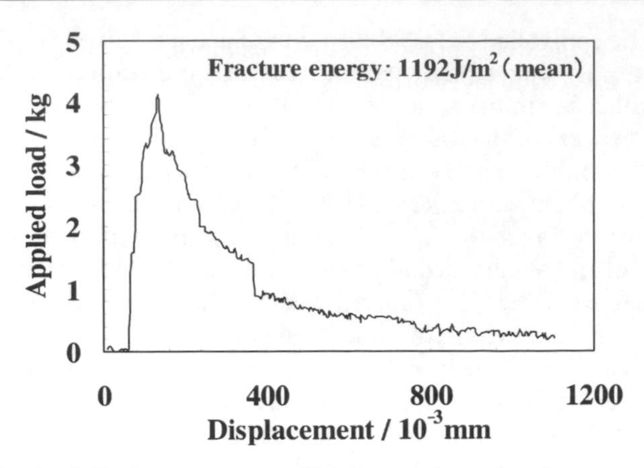

Fig. 22 Bending load-displacement curve of SA-Tyrannohex using chevron notch spacemen at room temperature

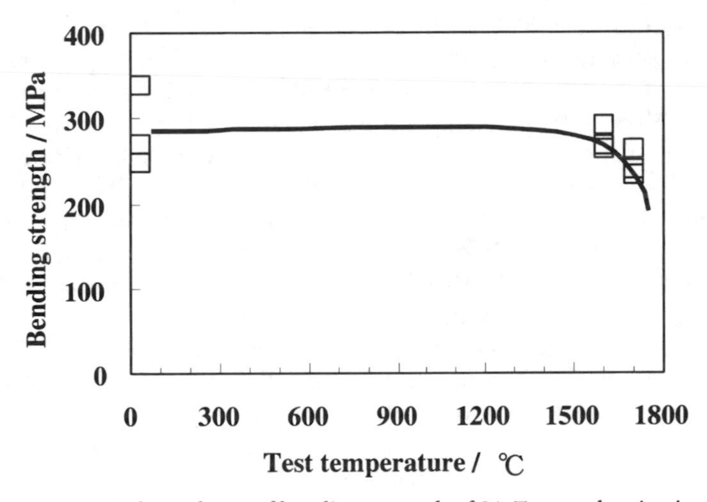

Fig. 23 Temperature-dependence of bending strength of SA-Tyrannohex in air

SiC/SiC is due to the change in the fiber property at high temperatures. However, the sintered SiC fiber, which consists of the same composition and interior structure as the SA-Tyrannohex, is very stable up to 2000 °C [2]. Moreover, the sintered SiC fiber shows negligible stress relaxation up to higher temperatures compared with other representative SiC fibers. Based on these findings, the high-temperature strength of the SA-Tyrannohex is attributed to the high-temperature properties of the fiber element.

Researchers have been developing SiC-CMCs in order to obtain an oxidation-resistant, tough thermostructural material. In general, a SiC-based material easily forms a protective oxide layer on its surface at high temperatures in air, leading to the well-known excellent oxidation resistance. The formation of the protective oxide layer proceeds in air according to the following reaction:

$$2SiC + 3O_2 = 2SiO_2 + 2CO$$

In this reaction, the oxygen diffusion through the oxide layer is the rate-determining step. However, at temperatures above 1600 °C in air, considerable vaporization of SiO or SiO_2 from the formed oxide layer begins to occur, so that a weight loss of the SiC-based material becomes conspicuous under the above conditions. Accordingly, as long as non-coated SiC-based material is used in air, the upper limit of temperature is at around 1600 °C. The SA-Tyrannohex, which showed no change even at 1900 °C in argon, also showed no marked weight loss up to 1600 °C in air, a temperature at which this material still retains its initial strength. From these findings, the SA-Tyrannohex is found to furnish with sufficient heat-resistance even in air.

The SA-Tyrannohex has potential for use in heat exchangers due to its relatively high thermal conductivity at temperatures above 1000 °C. Figure 24 shows the thermal conductivity of the SA-Tyrannohex in the direction through the thickness and the fiber direction along with other materials including representative SiC/SiC composite (CVI).

In general, the thermal conductivity of ceramics with strong covalent bonds is caused mainly by the transmission of phonons (lattice vibration). According to this theory [33], the desirable conditions for high thermal conductivity are as follows: (i) small mass of constituent atoms, (ii) strong bonding strength between the constituent atoms, (iii) short distance between neighboring atoms,

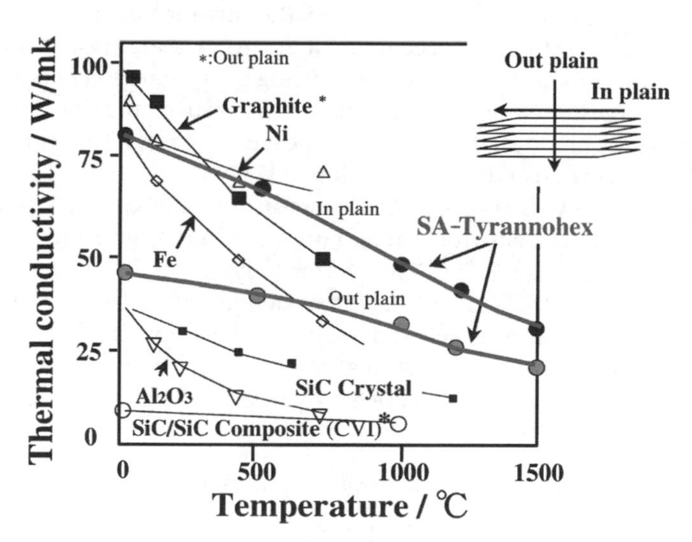

Fig. 24 Thermal conductivity of SA-Tyrannohex compared with other materials including metals

(iv) simple crystalline structure, and (v) high symmetry in lattice vibrations. Fundamentally, although the SiC crystal satisfies the above conditions, in the case of polycrystalline ceramics like the present SA-Tyrannohex, an ordered structure at the grain boundary is important for obtaining high thermal conductivity. As mentioned above, in the fiber element of the fiber-bonded ceramic, many SiC crystals of 0.1 to 0.4 μm in diameter are directly in contact with other similar crystals without any obvious intercrystalline phase. Furthermore, the fiber-bonded ceramic showed almost void-less structure (porosity, less than 1 vol.%) and very high fiber volume fraction (~100%). This structure accounts for the very high thermal conductivity of the SA-Tyrannohex compared with other representative SiC-CMCs. For example, the thermal conductivity of Nicalon-SiC/SiC (CVI) with a pyrolytic carbon interface is about 7 W/mK.

5 Strong Photocatalytic Fiber (Titania/Silica Fiber) Produced from Polycarbosilane

As mentioned above, many types of polymer-derived ceramic fibers have been developed using a polycarbosilane $(-SiH(CH_3)-CH_2-)_n$ as the starting material. Recently, using this base technology (precursor methods using a polycarbosilane), other types of ceramic fibers with functional surface layers with a nanometer-scale compositional gradient [7, 34] were produced. These fibers were produced from a polycarbosilane containing an excess amount of selected low-molecular-mass additives, which can be converted into functional ceramics by heat-treatment. Thermal treatment of the precursor fiber leads to controlled phase separation ("bleed out") of the low-molecular-mass additives from inside to outside of the precursor fibers. After that, subsequent calcination generates a functional surface layer during the production of bulk ceramic components. As the embodied functional material of the abovementioned process, a strong photocatalytic fiber composed of anatase-TiO_2 surface structure and silica core structure was developed [7].

Anatase-TiO_2 is well known as one of semiconductor catalysts, that exhibit a better photocatalytic activity by irradiation of a light whose energy is greater than the band gap (3.2 eV) [35]. The photocatalytic activity appears by irradiation of an ultraviolet (UV) light whose wavelength is shorter than 400 nm. The decomposition of harmful substances using the photocatalytic activity of anatase-TiO_2 has attracted a great deal of attention [36–39]. This effect is attributed to the generation of the strong oxidant ·OH radical according to the following reactions:

$$TiO_2 + h\nu \rightarrow e^- \text{ (electron)} + p^+ \text{ (hole) (excitation of } TiO_2 \text{ by light)} \quad (1)$$

$$p^+ \text{ (hole)} + OH^- \rightarrow \cdot OH \; (H_2O \rightarrow H^+ + OH^-) \quad (2)$$

Oxidation of harmful substances by the formed ·OH (3)

At present, most researches have been performed using a film or powder material. Of these, powder photocatalysts have some difficulties in practical use. For example, they have to be filtrated from treated water. Film photocatalysts cannot provide sufficient contact area with harmful substances. In order to avoid these problems, other types of research concerning fibrous photocatalysts have been conducted. However, up to the present, it has not been achieved to combine excellent photocatalytic activity and high fiber strength as long as sol-gel method was adopted. According to the new process, which will be introduced in this section, the development of a new photocatalytic, strong (2.5 GPa) and continuous fiber with small diameter (8 μm) was achieved. Namely, a type of titania-dispersed silica-based fiber with a sintered anatase-TiO_2 layer on the surface was developed. The surface layer comprised of nanoscale TiO_2 crystals (8 nm) was strongly sintered and exhibited excellent photocatalytic activity, which can lead to the efficient decomposition of harmful substances and bacterium contained in air and/or water by irradiation of UV light. In this section, the aforementioned photocatalytic fiber produced by the new in situ process and its actual applications are described. The photocatalytic fiber (titania/silica fiber) is produced by the following process. Polytitanocarbosilane containing an excess amount of titanium alkoxide was synthesized by the mild reaction of polycarbosilane $(\text{-SiH(CH}_3)\text{-CH}_2\text{-})_n$ (20 kg) with titanium(IV) tetra-*n*-butoxide (20 kg) at 220 °C in nitrogen atmosphere. The obtained precursor polymer was melt-spun at 150 °C continuously using melt-spinning equipment with a winding drum. The spun fiber, which contained excess amount of nonreacted titanium alkoxide, was pre-heat-treated at 100 °C and subsequently fired up to 1200 °C in air to obtain continuous transparent fiber (diameter: 8 μm). The appearance is shown in Fig. 25. In the initial stage of the pre-heat-treatment, effective bleeding of the excess amount of nonreacted titanium compound from the spun fiber occurred to form the surface layer containing large amount of titanium compounds. During the next firing process, the pre-heat-treated precursor fiber was converted into a titania-dispersed, silica-based fiber with a sintered TiO_2 layer on the surface according to the following reactions:

1. $(\text{-SiH(CH}_3)\text{-CH}_2\text{-})_n+4.5O_2=nSiO_2+3nH_2O+2nCO_2$
2. $Ti(OC_4H_9)_4+24O_2=TiO_2+18H_2O+16CO_2$

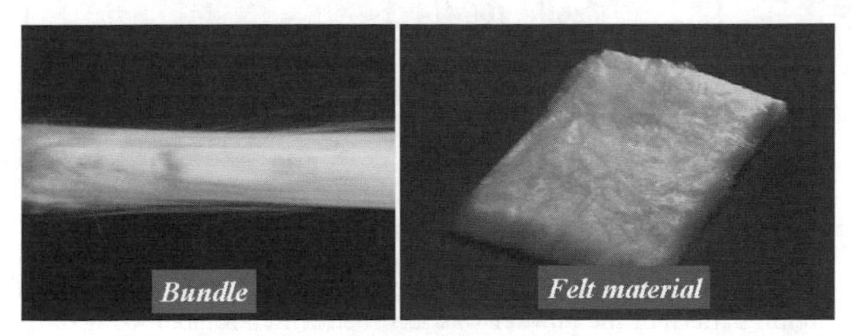

Fig. 25 The appearance of the photocatalytic fiber (diameter: 5~7 μm, strength: 2.5 GPa)

3. Formation of TiO_2 and SiO_2 from the reacted polycarbosilane
4. Densification and/or sintering of SiO_2 phase and TiO_2 phase

The fundamental concept of a new production process for the titania/silica fiber is shown in Fig. 26. The important feature of this method is that the surface titania layer of the fiber is not deposited on the substrate, but is formed during the production of the bulk structure. The gradient-like structure resulted in strong adhesion between the surface TiO_2 layer and the bulk material, which is different from the behavior of other coating layers formed on substrates by means of conventional methods. Accordingly, this TiO_2 surface layer definitely did not drop off after heat-cycling, washing or rubbing. This fiber con-

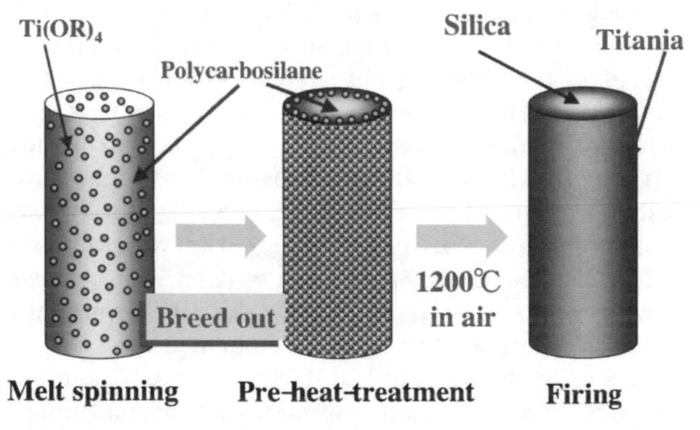

Fig. 26 The fundamental concept of a new production process for the TiO_2 fiber

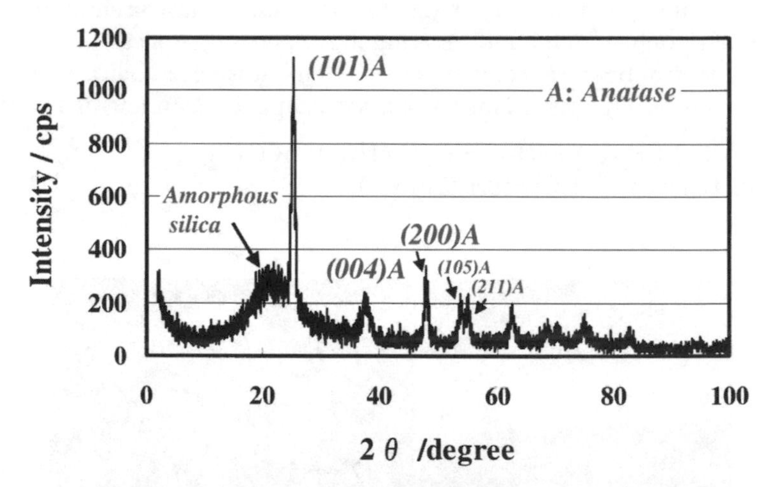

Fig. 27 X-ray diffraction pattern of our TiO_2 fiber. The TiO_2 fiber was pulverized, and the X-ray diffraction pattern of the powder was recorded with a Rigaku X-ray diffractometer with CuKα radiation with a nickel filter

tained 9.4% of titanium, 43.6% of silicon and 47% of oxygen, and was mainly composed of anatase-TiO_2 along with amorphous silica (Fig. 27). Although this fiber was fired at very high temperature (1200 °C), no obvious rutile phase could be observed. It is well known that anatase-TiO_2 converts to rutile at temperatures ranging from 700 °C to 1000 °C. In particular, pure nanocrystalline anatase easily converts to rutile at lower temperature (~500 °C) [40]. In the aforementioned new process, it is thought that the surrounding SiO_2 phase caused the stabilization of the anatase phase. At the interface between TiO_2 and SiO_2, atoms constructing TiO_2 are substituted into the tetrahedral SiO_2 lattice forming tetrahedral Ti sites [41]. The interaction between the tetrahedral SiO_2 species and the tetrahedral Ti sites in the anatase is thought to prevent the transformation to rutile.

As can be seen from Fig. 28, the surface of our TiO_2 fiber is densely covered with nanoscale anatase-TiO_2 particles (8 nm), which are strongly sintered with each other directly or through with amorphous silica phase (Fig. 28D). The thickness of the surface TiO_2 layer is approximately 100~200 nm (Fig. 28C). The tensile strength of this fiber measured by a single filament method was approximately 2.5 GPa. This mechanical strength is markedly superior to that of existing photo-catalytic TiO_2 fibers (<1 GPa), which were produced by a sol-gel method [42] or using polytitanosiloxanes [43]. The high strength of the titania/silica fiber is closely related to the dense structure without pores, which is caused by its higher firing temperature compared with previous TiO_2 fibers.

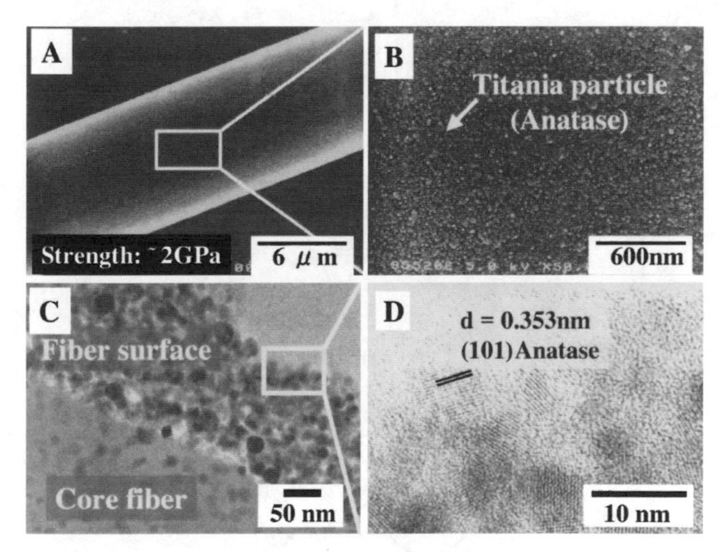

Fig. 28a–d SEM micrographs and TEM image of our TiO_2 fiber: **a, b** SEM micrographs of the surface (**b** – enlarged one of a part of the surface); **c, d** TEM image of our TiO_2 fiber (**c** – enlarged one of the surface layer). SEM micrographs and the TEM image were obtained with a Hitachi S-5000 operating at 5 kV, and a JEM 2010F operating at 200 kV, respectively

The photocatalytic activity of this titania/silica fiber was investigated using closed circulation equipment with UV light and a heater (Fig. 29). The felt material (0.3 g) of this fiber was placed in a quartz reactor. The thickness of this felt material was adjusted for UV light not to leak through the felt material to the reverse side. The ambient air including 250 ppm of acetaldehyde sealed in this equipment was initially circulated without UV light irradiation for 20 min in order to obtain adsorption equilibrium. This circulating gas was perfectly passed through the felt material of the titania/silica fiber. Subsequently, UV light (1 mW/cm^2) was irradiated at various temperatures with circulation (1 L/min). The changes in the concentration of acetaldehyde and CO_2 were measured every 10 min using gas chromatography. The result after 60 min is depicted in Fig. 30 along with that of a comparative study using silica fiber. Below 300 °C, both fibers markedly adsorbed acetaldehyde solely by circulation for 20 min. Adsorption was not observed at 400 °C for either fiber. It is considered that a significant decrease in the concentration at 500 °C was caused by thermal degradation. Regarding the adsorption and thermal degradation of acetaldehyde without UV light irradiation, no large differences were observed between two types of fibers. After the circulation for 20 min, UV light irradiation was started. In the system using the titania/silica fiber, an abrupt decrease in the concentration of acetaldehyde was observed, accompanied by the formation of CO_2 gas. On the other hand, in the system using silica fiber, such a photocatalytic decomposition of acetaldehyde did not occur.

The photocatalytic effect of TiO_2 has been known for a long time. To date, most researches concerning this effect have been performed using powder or film ma-

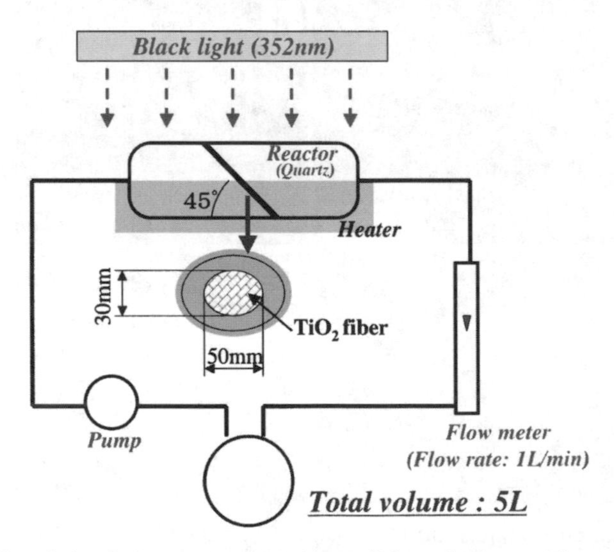

Fig. 29 The closed circulation equipment with UV light and a heater. The reactor, in which the felt material (0.3 g) of our TiO_2 fiber was placed, was made of quartz. Black light was used as a light source of UV light

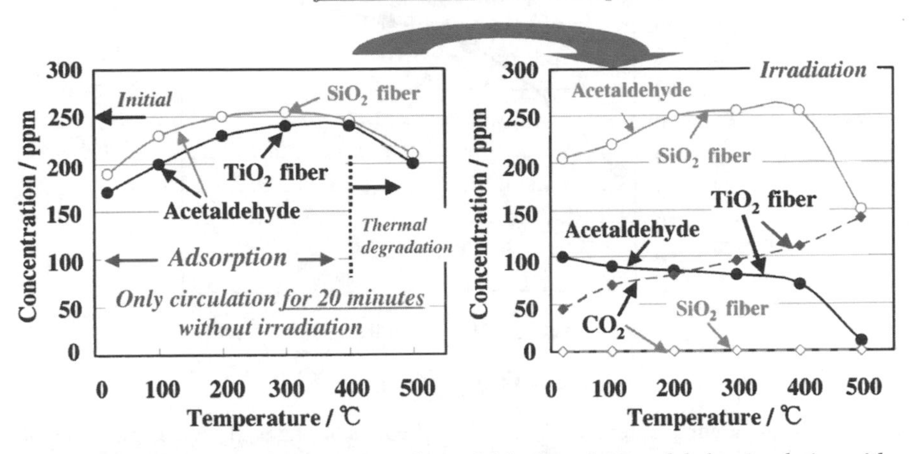

Fig. 30 Changes in the concentrations of acetaldehyde and CO_2 solely by circulation without irradiation and after subsequent irradiation of UV light at various temperatures. The felt material (0.3 g) was placed at the center of the quartz reactor. The initial concentration of acetaldehyde was 250 ppm.

terial of TiO_2. In the case of powder material, the catalyst filtration step after the photocatalytic decomposition of the organic material is necessary. Besides, a film material could not achieve a large contact area compared with the felt material of the thin titania/silica fiber. The strong thin titania/silica fiber also has good fabrication ability, and thus can be used widely – for example, in wastewater, in combustion gas, in ambient air containing harmful materials, and so forth.

It is well known that the catalytic effect of TiO_2 is attributed to the generation of a strong oxidant, hydroxyl radicals [44]. Following this theory, the quantum efficiency of the felt material prepared with the titania/silica fiber was calculated from the aforementioned result. In this case, if the number of molecules is significantly larger than the number of photon, acetaldehyde is oxidized to CH_3COOH as follows:

$$CH_3CHO + H_2O + 2H^+ = CH_3COOH + 2H^+$$

*Semiconductor (such as TiO_2) + hν(energy of photon)=e^- (electron)+h^+ (hole)

In this case, the apparent quantum efficiency (QE) of the felt material prepared from the titania/silica fiber is calculated by the following equation:

$$QE = 2 \times (\text{Number of decomposed molecules})\ (\text{Number of incident photons})$$
$$\text{Wavelength} = 352 \times 10^{-9}\ \text{m, Intensity of UV light} = 1\ \text{mW/cm}^2 = 10\ \text{J/s/m}^2$$
$$\text{Actual irradiation area} = 8.33 \times 10^{-4}\ \text{m}^2$$

The calculation result using these values is presented in Fig. 31. As can be seen from this figure, the felt material of the titania/silica fiber showed an extremely

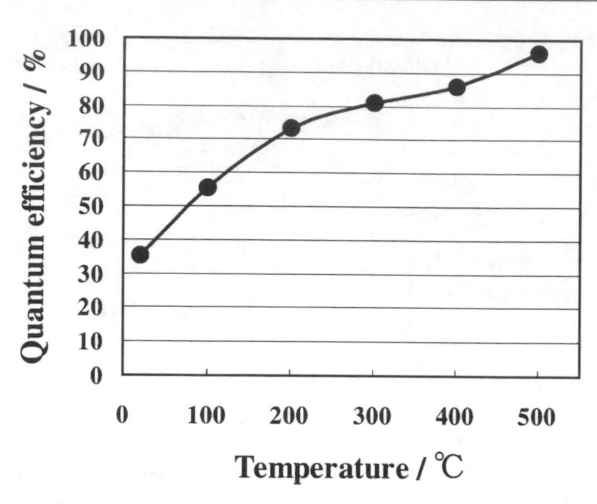

Fig. 31 The apparent quantum efficiency of the titania/silica fiber. The wavelength and the intensity of the light at the surface of the felt material were 352 nm and 1 mW/cm^2, respectively. Actual irradiation area perpendicular to the ray of light was 8.33×10^{-4} m^2

high QE value even at room temperature (37%). Furthermore, at higher temperatures over 200 °C, an excellent QE value (over 70%) was obtained. However, actual QE values are thought to be much higher than those values, because the formation of CO_2 was obviously observed. It is believed that these higher values at high temperatures are attributable to the evaporation of the formed acetic acid adsorbed on the fiber surface. These excellent QE values could be realized by the dense existence of nanoscale anatase-TiO_2 crystals (8 nm) on the surface. These nanoscale crystals are considered to facilitate the diffusion of excited electrons and holes toward the surface before their recombination.

Lastly, the coliform-sterilization ability of this fiber was confirmed as follows. The titania/silica fiber (0.2 g) was placed in wastewater (20 ml) containing coliform of 2×10^6 ml^{-1}. Irradiation of UV light (2 mW/cm^2) was performed at room temperature, and then a small amount of the wastewater was extracted every 1 h. After cultivation using the extracted water, the amount of active coliform was calculated from the number of formed colonies. In this experiment, all of the coliform included in the wastewater was completely sterilized within 3 h. In this experiment, using the desirable fiber covered with very fine titania crystals (8 nm), all of the coliform in the wastewater was completely sterilized within 5 h accompanied by the generation of CO_2. In the comparative study, using undesirable fibers covered with large titania crystals (9~11 nm), sterilization of the coliform was markedly slow (Fig. 32). From the results, the size of the titania crystal is found to be closely related to the photo-catalytic activity. It is assumed that, in the case of large crystals, the recombination (inactivation) of the hole and excited electron generated by UV irradiation easily occurs (Fig. 33). In order to suppress the recombination and obtain the good photo-catalytic activity, the creation of

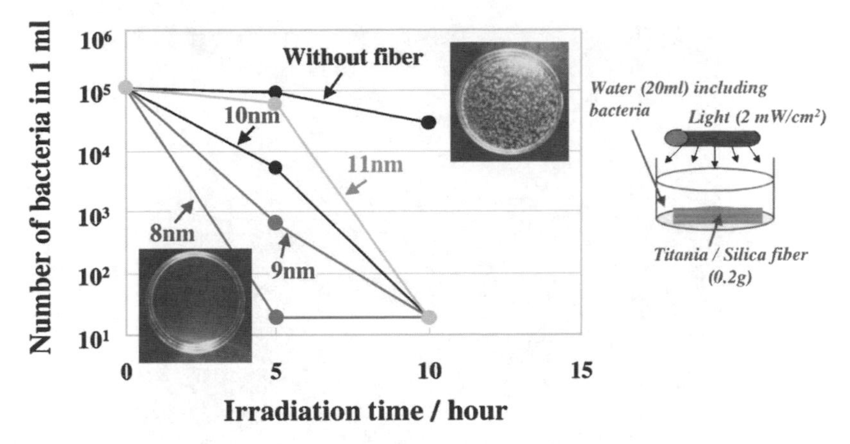

Fig. 32 The results of extinction activity of coliform using the photo-catalytic fiber with UV irradiation

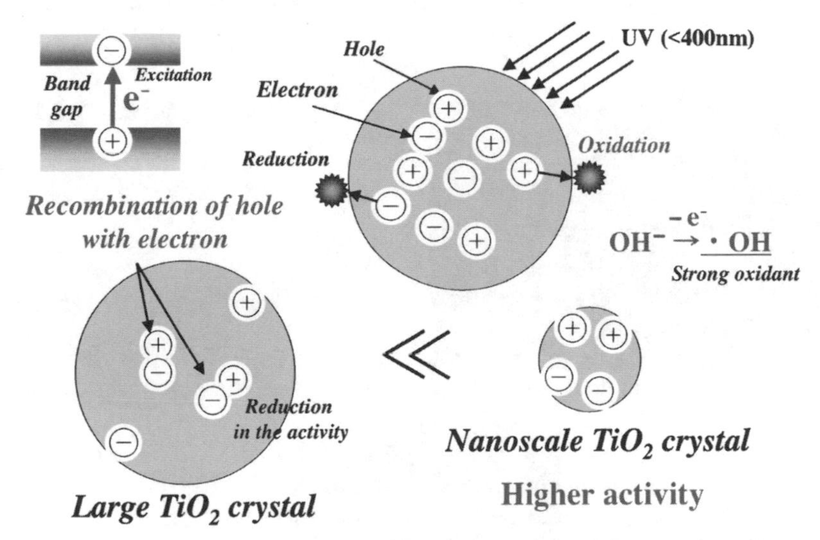

Fig. 33 The relationship between the photo-catalytic activity and the size of a titania crystal

the smaller titania crystals is very important. The new process described in this section is very desirable for controlling the size of fine crystals, because both the bleed-out phenomenon of the low-molecular-mass additive and the crystallization of the functional material proceed competitively. Industrial application of the photocatalytic fiber with a gradient titania layer is shown as follows.

A circulation purifier for pollutants (Fig. 34) was developed using the felt material made of the aforementioned photo-catalytic fiber. This is a very simple purifier with a module composed of the cone-shaped felt material (made of the photo-catalytic fiber) and UV lamp. Purification of the bath water of a

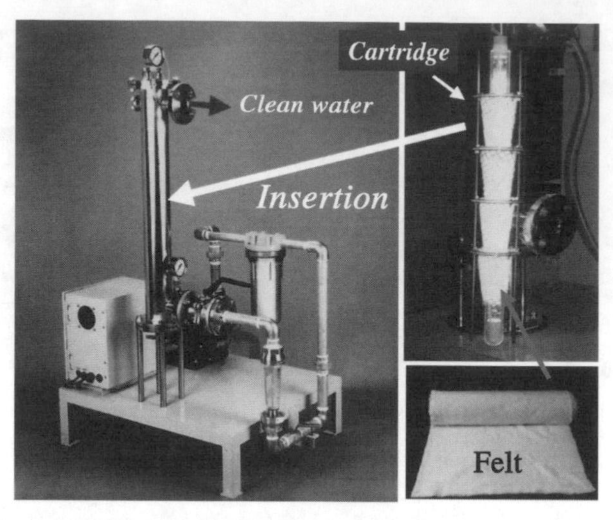

Fig. 34 Circulation purifier for pollutants using photo-catalytic fiber with a UV lamp

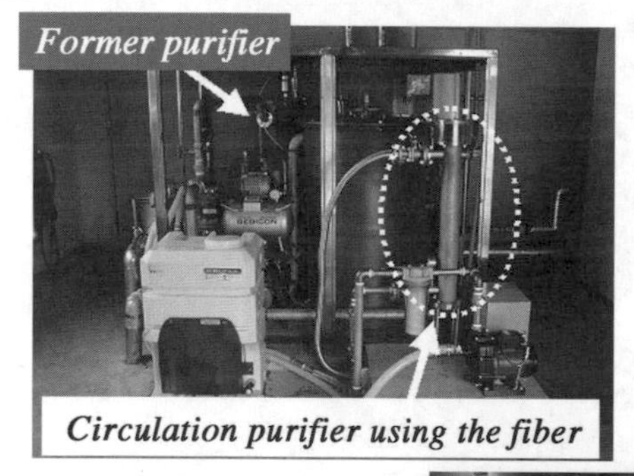

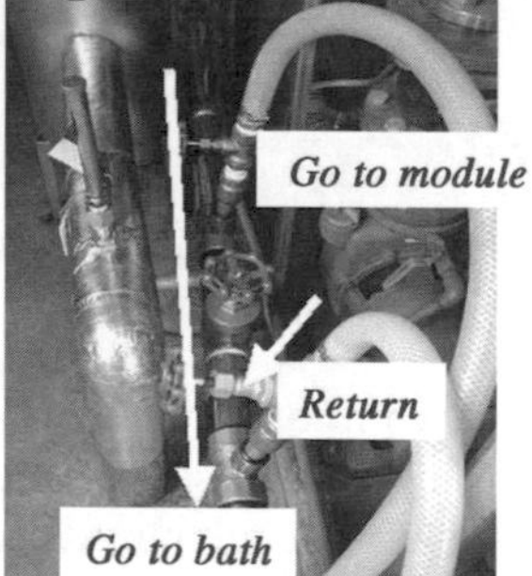

Fig. 35 Purification test of the water of a circulation bath system using the circulation purifier (bath size: 8 m^3, users: 200 persons/day, circulation of water: 3 m^3/h in the water-purifier, sampling: directly from the bath)

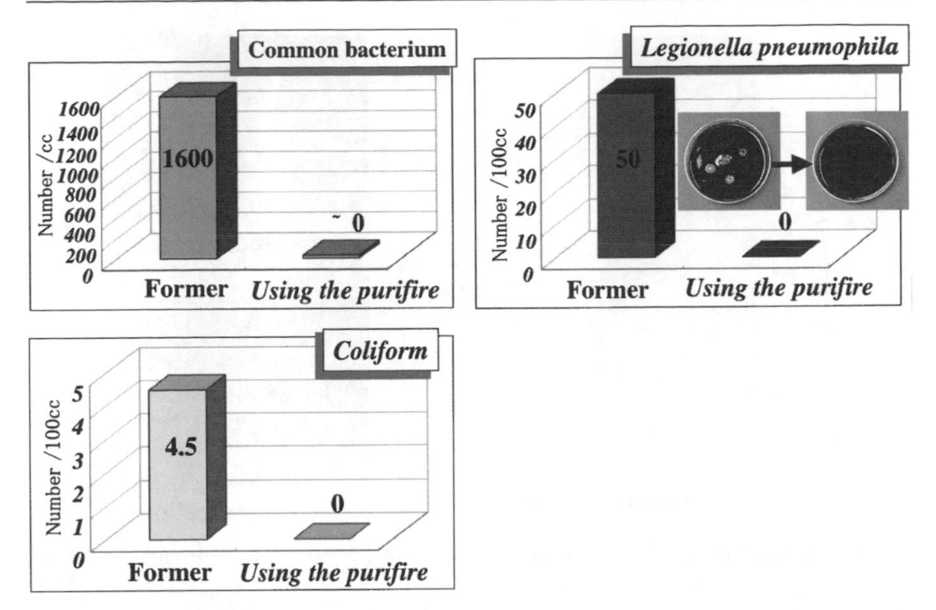

Fig. 36 The results of the purification test using the circulation purifier (bath size: 8 m^3, users: 200 persons/day, circulation of water: 3 m^3/h in the water-purifier, sampling: directly from the bath)

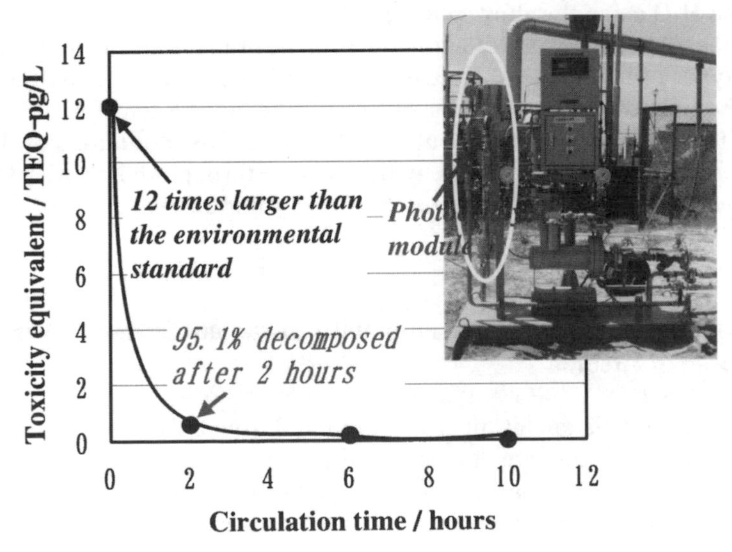

Fig. 37 Decomposition of dioxin using the circulation purifier (photo-intensity: 10 mW/cm^2, water quantity: 120 L, circulation: 400 L/h)

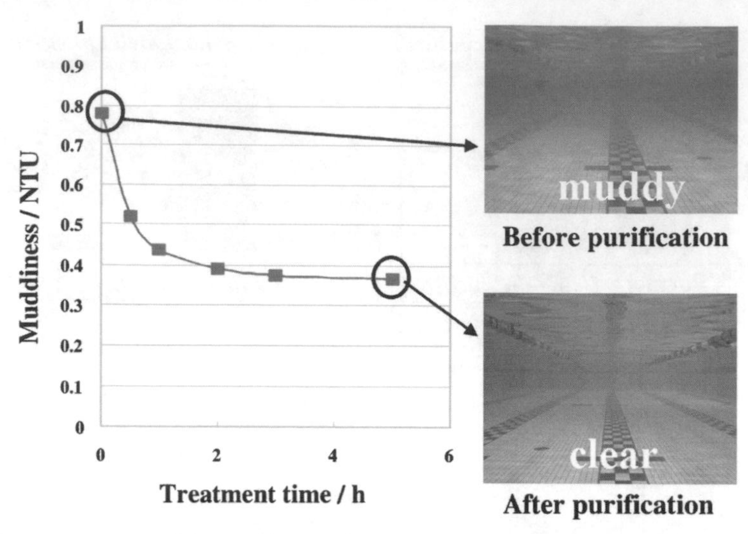

Fig. 38 Improvement of the muddiness of the pool water by the passage through the purifier

circulation bath system was performed using the above purifier (Fig. 35). Many bacteria (common bacterium, *Legionera pneumophila* and coliform), which existed in the bath water before the purification, were perfectly decomposed into CO_2 and H_2O using the above purifier (Fig. 36).

Furthermore, the photo-catalytic fiber can be used for the purification of the many types of wastewater. Figure 37 shows the result regarding a decomposition of dioxin contained in a wastewater. In this case, 95.1% of the dioxin was found to be decomposed after only 2 h. Moreover, the muddiness of the pool water was remarkably improved by the passage through the purifier (Fig. 38) along with a decrease in the organic filth and chloramines.

6 Oxidation-Resistant SiC-Based Fiber with a Gradient Surface Layer Composed of Zirconia

The oxidation-resistant SiC fiber was prepared from polycarbosilane containing $Zr(OC_4H_9)_4$ by the same process as that used for the aforementioned titania/silica fiber, except that the calcination was performed in Ar atmosphere at 1400 °C. In this case, the polycarbosilane and $Zr(OC_4H_9)_4$ were effectively converted into SiC-based bulk ceramic and zirconium oxide (cubic zirconia). Before the conversion, bleed-out of the zirconium compound proceeded effectively. AES depth analysis of the fiber surface showed an increase in the concentration of zirconium towards the surface. This construction was confirmed by the TEM image of the cross-section near the fiber surface. This indicates the direct production of a SiC-based fiber covered with a ZrO_2 surface layer, which

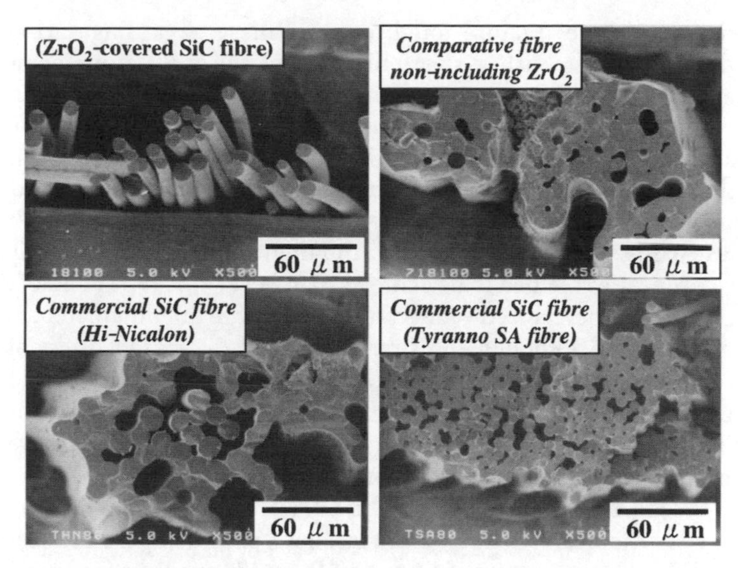

Fig. 39 Alkali resistance of the ZrO_2/SiC fiber with comparative results

has a gradient-like composition towards the surface. In general, amorphous fibers covered with ceramic crystal do not show high strength. However, this fiber showed relatively high strength (2.5 GPa) compared with other SiC fiber (2.1 GPa) coated with zirconia nanocrystals by means of the sol-gel method. The initial strength of the SiC fiber used for the comparative study was 3.1 GPa. The ZrO_2 surface layer, a basic oxide material, can provide better alkali resistance for SiC ceramics. In order to confirm the better alkali resistance for the aforementioned ZrO_2-covered SiC fiber, the following experiment was performed. The fiber material was immersed for 15 min in deionized water saturated with potassium acetate and then annealed at 800 °C for 100 h in air after drying. Comparative studies were conducted using the SiC-based fiber prepared from polycarbosilane, which did not contain zirconium(IV) butoxide, as well as commercial SiC fibers, namely, Hi-Nicalon and an alkali-resistant sintered SiC fiber (Tyranno SA fiber). Figure 39 shows the fractured surfaces of the tested fiber bundles, obtained using field-emission scanning electron microscopy (FE-SEM). As can be seen from the micrographs, only the above ZrO_2-covered SiC fiber retained its intact fibrous shape, whereas the other SiC fibers were extensively oxidized and then bonded together.

7 Summary and Prospect

Many types of inorganic fibers have been developed and commercialized over the last 30 to 40 years. The main target was to develop composite materials with lightweight and high fracture toughness. Of these, regarding carbon fiber, which has already established a very big market, the application technologies (for example: composite technology and coating techniques) are mainly fully developed now. Small-diameter oxide fibers based on polycrystalline alumina/silica retain the desired mechanical properties up to 1200 °C but much above this temperature they show remarkable structural changes and loss of mechanical strength. In order to increase the usable temperature of fibers, the two-phase oxide system was developed, and then an improvement in high-temperature properties seemed to be achieved. However, this type of eutectic fiber has not been commercialized yet because of the large fiber diameter and lower production ability. Although the first developed SiC-based fibers also had limitations in their usable temperature, around 1200 °C, finally several types of excellent heat-resistant SiC-polycrystalline fibers were synthesized from a polycarbosilane. These types of SiC-polycrystalline fibers have achieved both excellent heat-resistance and oxidation resistance.

All of the aforementioned inorganic fibers were developed in order to obtain improved mechanical strength and excellent high-temperature-properties. However, recently a new type of ceramic fiber with excellent function (for example, photocatalytic activity) was developed based on the aforementioned production technology of SiC-based fibers using a polycarbosilane. This fiber also has both high strength and heat-resistance. In the following ten years, many types of functional ceramic fibers with not only mechanical strength and heat-resistance but also excellent functions will surely be developed and commercialized (Fig. 40).

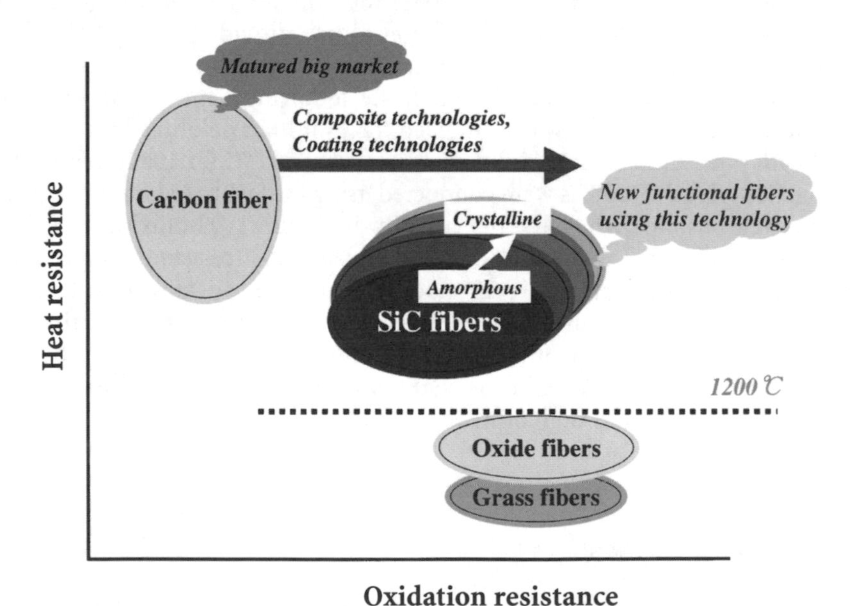

Fig. 40 Prospect for the future inorganic fibers

References

1. Boakye E, Hay RS, Petry MD (1999) J Am Ceram Soc 82:2321
2. Ishikawa T, Kohtoku Y, Kumagawa K, Yamamura T, Nagasawa T (1998) Nature 391:773
3. Ishikawa T (2000) Ann Chim Sci Mat 25:517
4. Ichikawa H (2000) Ann Chim Sci Mat 25:523
5. Yun HM, DiCarlo JA (2003) Thermomech Properties 40:15
6. Yang W, Araki H, Kohyama A, Busabok C, Hu Q, Suzuki H, Noda T (2003) Mater Trans 44:1797
7. Ishikawa T, Yamaoka H, Harada Y, Fujii T, Nagasawa T (2002) Nature 416:64
8. Effinger MR (2000) Adv Mater Process 157(6):69
9. Richard JP (1998) Sci Technol Carbon 2:733
10. Davis JB, Marshall DB, Oka KS, Housley RM, Morgan PED (1999) Composites A(30): 483
11. Chawla KK, Coffin C, Xu ZR (2000) Int Mater Rev 45:165
12. Takeda M, Sakamoto J, Saeki A, Ichikawa H (1996) Ceram Eng Sci Proc 17:35
13. Urano A, Sakamoto J, Takeda M, Imai Y (1998) Ceram Eng Sci Proc 19(3):55
14. Kumagawa K, Yamaoka H, Shibuya M, Yamamura T (1997) Ceram Eng Sci Proc 18: 113
15. Xu Y, Zangvil, Lipowitz J, Rabe JA, Zank GA (1993) J Am Ceram Soc 76:3034
16. Kumagawa K, Yamaoka H, Shibuya M, Yamamura T (1998) Ceram Eng Sci Proc 19A:65
17. Koyama A, Seki M, Abe K, Muroga T, Matsui H, Jitsukawa S, Matsuda S (2000) J Nucl Mater 283–287:20
18. Hinoki T, Katoh Y, Koyama A (2002) Mater Trans 43:617
19. Jacques S, Guette A, Langlais F, Naslain R (1997) J Mater Sci 32:983
20. Sun EY, Nutt SR, Brennan JJ (1996) J Am Ceram Soc 79(6):1521
21. Masaki S, Moriya K (1996) Ceram Trans 58:187
22. Parthasarathy TA, Mah T, Folsom CA, Katz AP (1995) J Am Ceram Soc 78:1992
23. Yamamura T, Ishikawa T, Shibuya M (1992) Japan Patent 1699835
24. Tanaka H, Inomata Y, Hara K, Hasegawa H (1985) J Mater Sci Lett 4:315
25. Bocker W, Landfermann H, Hausner H (1979) Powder Metall Int 11:83
26. Alliegro RA, Coffin LB, Tinklepaugh JR (1956) J Am Ceram Soc 39:386
27. Izeki T (1985) Handbook of corrosion resistance of ceramics. Kyoritsu Syuppan, Japan, pp 61–67
28. Ishikawa T, Kajii S, Matsunaga K, Hogami T, Kohtoku Y, Nagasawa T (1998) Science 282:1295
29. Takeda M, Imai Y, Ichikawa H, Kagawa Y, Iba H, Kakisawa H (1997) Ceram Eng Sci Proc 18:779
30. Morscher GN (1997) Ceram Eng Sci Proc 18:737
31. Ifflander K, Gualco GC (1997) Ceram Eng Sci Proc 18:625
32. Tanaka M, Imai Y, Ichikawa H, Ishikawa T, Kasai N, Seguchi T, Okamura K (1993) Ceram Eng Sci Proc 14:540
33. Kingery WD, Brown HK, Uhlman DR (1976) Introduction to ceramics, 2nd edn. Wiley, New York, p 583
34. Ishikawa T, Harada Y, Hayashi H, Kajii S (2003) US Patent 6,541,416 B2 (Foreign Application Priority Date: June 13, 2000)
35. Sopyan I, Murasawa S, Hashimoto K, Fujishima A (1994) Chem Lett 723
36. Matsuda A, Kotani Y, Kogure T, Tatsumisago M, Minami T (2000) J Am Ceram Soc 83:229

37. Park DR, Zhang J, Ikeue K, Yamashita H, Anpo M (1999) J Catal 185:114
38. Takeda N, Ohtani M, Torimoto T, Kuwabata S, Yoneyama H (1997) J Phys Chem B 101:2644
39. Chan CK, Porter JF, Li YG, Guo W, Chan CM (1999) J Am Ceram Soc 82:566
40. Gouma PI, Dutta PK, Mills MJ (1999) Nanostruct Mater 11:1231
41. Anderson C, Bard AJ (1997) J Phys Chem B 101:2611
42. Koike H, Oki Y, Takeuchi Y (1999) Mater Res Soc Sym Pro 549:141
43. Gunji T, Sopyan I, Abe Y (1991) J Polym Sci Part A Polym Chem 32:3133
44. Noda H, Oikawa K, Ohya-Nishiguchi H, Kamada H (1994) Bull Chem Soc Jpn 67:2031

Author Index Volumes 101–178

Author Index Volumes 1–100 see Volume 100

Subject Index